T0214177

Communications in Computer and Information Science 927

Commenced Publication in 2007
Founding and Former Series Editors:
Phoebe Chen, Alfredo Cuzzocrea, Xiaoyong Du, Orhun Kara, Ting Liu,
Dominik Ślęzak, and Xiaokang Yang

More information about this series at http://www.springer.com/series/7899

Davide Buscaldi · Aldo Gangemi
Diego Reforgiato Recupero (Eds.)

Semantic Web Challenges

5th SemWebEval Challenge at ESWC 2018
Heraklion, Greece, June 3–7, 2018
Revised Selected Papers

 Springer

Editors
Davide Buscaldi (iD)
Laboratoire d'Informatique de Paris-Nord
Paris 13 University
Villetaneuse, France

Aldo Gangemi (iD)
University of Bologna
Bologna, Italy

Diego Reforgiato Recupero
Department of Mathematics and Computer
 Science
University of Cagliari
Cagliari, Italy

ISSN 1865-0929   ISSN 1865-0937   (electronic)
Communications in Computer and Information Science
ISBN 978-3-030-00071-4   ISBN 978-3-030-00072-1   (eBook)
https://doi.org/10.1007/978-3-030-00072-1

Library of Congress Control Number: 2017956784

This Springer imprint is published by the registered company Springer Nature Switzerland AG
The registered company address is: Gewerbestrasse 11, 6330 Cham, Switzerland

Preface

Reproducible, evaluable, and comparable scientific research is based on common benchmarks, established evaluation procedures, comparable tasks, and public datasets. Open challenges are now a key scientific element of various conferences aimed at specific research communities. The Semantic Web research community is no exception, and the challenge track is an important element for assessing the current state of the art and for fostering the systematic comparison of contributions. Following the success of the previous years, reflected by the high number of high-quality submissions, we organized the fifth edition of the Semantic Web Challenges as an official track of the ESWC 2018 conference (held in Heraklion, Greece, during June 3–7, 2018), one of the most important international scientific events for the Semantic Web research community. The purpose of challenges is to validate the maturity of the state of the art in tasks common to the Semantic Web community and adjacent academic communities in a controlled setting of rigorous evaluation, thereby providing sound benchmarks, datasets, and evaluation approaches, which contribute to the advancement of the state of the art. This fifth edition included four challenges: Open Knowledge Extraction (OKE 2018), Semantic Sentiment Analysis (SSA 2018), Scalable Question Answering Challenge over Linked Data (SQA2018), and The Mighty Storage Challenge (MOCHA 2018). A total of 17 teams competed in the different challenges. The event attracted participants from across the conference, with a high attendance for all challenge-related activities during ESWC 2018. This included the dedicated conference track and participation of challenge candidates during the ESWC poster and demo session. The very positive feedback and resonance suggests that the ESWC challenges were a central contribution to the ESWC 2018 program.

Therefore, all the approaches and systems competing at the Semantic Web Challenge 2018 are included in this book. The book also contains the detailed description of each of the four challenges, their tasks, the evaluation procedures and the links to the scripts released to compute the evaluations, the related datasets and the URL from which to download them. The book therefore offers the community a picture of the advancements in each challenge domain releasing the material for the replication of the results.

Each chapter starts with an introductory section by the challenge chairs that contains a detailed description of their challenge tasks, the evaluation procedure, and associated datasets, and peer-reviewed descriptions of the participants' approach, comparisons, tools, and results.

The 2018 edition of the Semantic Web Challenge was truly successful and especially thanks to the hard work of the chairs of each challenge for the organization of their challenges. We also thank all the challenge participants, who provided novel systems showing great advancements in the technology they employed, which we are glad to share with the entire scientific community through this book. Moreover, we thank the organizers of the main conference, ESWC 2018, who paved the way with researchers and professionals from industry for a cross-pollination of ideas among them. Last but not least, we would like to thank Springer for having provided several awards to the winners of the different tasks in each challenge and for promoting the Semantic Web Challenge to increase its audience and coverage.

June 2018   Davide Buscaldi
                                                         Aldo Gangemi
                                            Diego Reforgiato Recupero

Organization

Organizing Committee (ESWC)

General Chair

Aldo Gangemi — University of Bologna and ISTC-CNR, Italy

Research Track

Roberto Navigli — Sapienza University of Rome, Italy
Maria-Esther Vidal — Leibniz Information Centre for Science and
    Technology University Library, Germany,
    and Universidad Simon Bolivar, Venezuela

Resource Track

Pascal Hitzler — Wright State University, USA
Raphael Troncy — Eurecom, France

In-Use Track

Laura Hollink — Centrum Wiskunde and Informatica, The Netherlands
Anna Tordai — Elsevier, The Netherlands

Workshop and Tutorials

Heiko Paulheim — University of Mannheim, Germany
Jeff Pan — University of Aberdeen, UK

Poster and Demo

Anna Lisa Gentile — IBM Research Almaden, USA
Andrea Nuzzolese — ISTC-CNR, Italy

Challenge Chairs

Davide Buscaldi — Université Paris 13, Villetaneuse, France
Diego Reforgiato Recupero — University of Cagliari, Italy

PhD Symposium

Sebastian Rudolf — Technische Universität Dresden, Germany

EU Project Networking

Maria Maleshkova — Karlsruhe Institute of Technology, Germany

Industry Session

Andrea Conte (Senior   Reply Spa, Italy
  Manager)

Sponsoring

York Sure-Vetter   Karlsruhe Institute of Technology, Germany

Publicity

Mehwish Alam   ISTC-CNR, Italy

Open Conference Data

Sarven Capadisli   University of Bonn, Germany
Silvio Peroni   University of Bologna, Italy

Web Presence

Venislav Georgiev   STI, Austria

Treasurer

Dieter Fensel   STI, Austria

Challenges Organization

Challenge Chairs

Davide Buscaldi   Université Paris 13, Villetaneuse, France
Diego Reforgiato Recupero   University of Cagliari, Italy

MOCHA2018: The Mighty Storage Challenge

Kleanthi Georgala   University of Leipzig, Germany
Mirko Spasić   OpenLink Software, UK
Milos Jovanovik   OpenLink Software, UK
Vassilis Papakonstantinou   Institute of Computer Science-FORTH, Greece
Claus Stadler   University of Leipzig, Germany
Michael Röder   University of Leipzig, Germany
Axel-Cyrille Ngonga   Institute for Applied Informatics, Germany
  Ngomo

Open Knowledge Extraction Challenge 2018

René Speck   University of Leipzig, Germany
Michael Röder   University of Leipzig, Germany
Felix Conrads   Paderbon University, Germany
Hyndavi Rebba   Paderbon University, Germany
Catherine Camilla Romiyo   Paderbon University, Germany

Gurudevi Salakki	Paderbon University, Germany
Rutuja Suryawanshi	Paderbon University, Germany
Danish Ahmed	Paderbon University, Germany
Nikit Srivastava	Paderbon University, Germany
Mohit Mahajan	Paderbon University, Germany
Axel-Cyrille Ngonga Ngomo	Institute for Applied Informatics, Germany

The Scalable Question Answering Over Linked Data (SQA) Challenge 2018

Giulio Napolitano	Fraunhofer-Institute IAIS, Germany
Ricardo Usbeck	University of Leipzig, Germany
Axel-Cyrille Ngonga Ngomo	Institute for Applied Informatics, Germany

Semantic Sentiment Analysis Challenge 2018

| Mauro Dragoni | Fondazione Bruno Kessler, Italy |
| Erik Cambria | Nanyang Technological University, Singapore |

Contents

Semantic Sentiment Analysis Challenge

The Mighty Storage Challenge

MOCHA2018: The Mighty Storage Challenge at ESWC 2018

Kleanthi Georgala[1], Mirko Spasić[2(✉)], Milos Jovanovik[2],
Vassilis Papakonstantinou[3], Claus Stadler[1], Michael Röder[4,5],
and Axel-Cyrille Ngonga Ngomo[4,5]

[1] AKSW Research Group, University of Leipzig, Goerdellering 9,
04109 Leipzig, Germany
{georgala,cstadler}@informatik.uni-leipzig.de
[2] OpenLink Software, London, United Kingdom
{mspasic,mjovanovik}@openlinksw.com
[3] Institute of Computer Science-FORTH, Heraklion, Greece
papv@ics.forth.gr
[4] Paderborn University, DICE Group, Warburger Str. 100,
33098 Paderborn, Germany
{michael.roeder,axel.ngonga}@uni-paderborn.de
[5] Institute for Applied Informatics (InfAI) e.V., AKSW Group,
Hainstr. 11, 04109 Leipzig, Germany

Abstract. The aim of the Mighty Storage Challenge (MOCHA) at ESWC 2018 was to test the performance of solutions for SPARQL processing in aspects that are relevant for modern applications. These include ingesting data, answering queries on large datasets and serving as backend for applications driven by Linked Data. The challenge tested the systems against data derived from real applications and with realistic loads. An emphasis was put on dealing with data in form of streams or updates.

1 Introduction

Triple stores and similar solutions are the backbone of most applications based on Linked Data. Hence, devising systems that achieve an acceptable performance on real datasets and real loads is of central importance for the practical applicability of Semantic Web technologies. This need is emphasized further by the constant growth of the Linked Data Web in velocity and volume [1], which increases the need for storage solutions to ingest and store large streams of data, perform queries on this data efficiently and enable high performance in tasks such as interactive querying scenarios, the analysis of industry 4.0 data and faceted browsing through large-scale RDF datasets. The lack of comparable results on the performance on storage solutions for the variety of tasks which demand time-efficient storage solutions was the main motivation behind this challenge. Our main aims while designing the challenge were to (1) provide

© Springer Nature Switzerland AG 2018
D. Buscaldi et al. (Eds.): SemWebEval 2018, CCIS 927, pp. 3–16, 2018.
https://doi.org/10.1007/978-3-030-00072-1_1

objective measures for how well current systems (including 2 commercial systems, which have already expressed their desire to participate) perform on real tasks of industrial relevance and (2) detect bottlenecks of existing systems to further their development towards practical usage.

2   The MOCHA Challenge

2.1   Overview

MOCHA2018 took place in conjunction with the 15th European Semantic Web Conference (ESWC 2018), 3rd-7th June 2018, Heraklion, Crete, Greece, for the second consecutive year[1] As last year, the aim of the Mighty Storage Challenge was to test the performance of solutions for SPARQL processing in aspects that are relevant for modern applications. These include ingesting data, answering queries on large datasets and serving as backend for applications driven by Linked Data. The proposed challenge will test the systems against data derived from real applications and with realistic loads. An emphasis will be put on dealing with changing data in form of streams or updates. We designed the challenge to encompass the following tasks:

1. Sensor Streams Benchmark, that measured how well systems can ingest streams of RDF data.
2. Data Storage Benchmark, that measured how data stores perform with different types of queries.
3. Versioning RDF Data Benchmark, that measured how well versioning and archiving systems for Linked Data perform when they store multiple versions of large data sets. This task is introduced for the first time this year.
4. Faceted Browsing Benchmark, that measured for how well solutions support applications that need browsing through large data sets.

2.2   Tasks

Task 1: Sensor Streams Benchmark. The aim of this task was to measure the performance of SPARQL query processing systems when faced with streams of data from industrial machinery in terms of efficiency and completeness. For this task, we developed ODIN (StOrage and Data Insertion beNchmark), that was designed to test the abilities of tripe stores to store and retrieve streamed data. Our goal was to measure the performance of triple stores by storing and retrieving RDF data, considering two main choke points: (1) scalability (Data volume): Number of triples per stream and number of streams and (2) time complexity (Data velocity): Number of triples per second. The input data for this task consists of data derived from mimicking algorithms trained on real industrial datasets. Each training dataset included RDF triples generated within a predefined period of time (e.g., a production cycle). Each event (e.g., each sensor

---

[1] https://2018.eswc-conferences.org/.

measurement or tweet) had a timestamp that indicates when it was generated. Data was generated using data agents in form of distributed threads. An agent is a data generator who is responsible for inserting its assigned set of triples into a triple store, using SPARQL INSERT queries. Each agent emulated a dataset that covered the duration of the benchmark. All agents operated in parallel and were independent of each other. The insertion of a triple was based on its generation timestamp. SPARQL SELECT queries were used to check when the system completed the processing of the particular triples. To emulate the ingestion of streaming RDF triples produced within large time periods within a shorter time frame, we used a time dilatation factor that allowed rescaling data inserts to shorter time frames.

Task 2: Data Storage Benchmark (DSB). This task consists of an RDF benchmark that measures how data storage solutions perform with interactive read SPARQL queries. Running the queries is accompanied with high insert rate of the data (SPARQL INSERT queries), in order to mimic realistic application scenarios where READ and WRITE operations are bundled together. Typical bulk loading scenarios are supported. The queries and query mixes are designed to stress the system under test in different choke-point areas, while being credible and realistic. The benchmark is based on and extends the Social Network Benchmark (SNB), developed under the auspices of the Linked Data Benchmark Council (LDBC)[2]. LDBC introduced a new choke-point driven methodology for developing benchmark workloads, which combines user input with input from expert systems architects [2]. The dataset generator developed for the previously mentioned benchmark is modified in DSB in order to produce synthetic RDF datasets available in different sizes, but more realistic and more RDF-like. The structuredness of the dataset is in line with real-world RDF datasets unlike the LDBC Social Network Benchmark dataset, which is designed to be more generic and very well structured. For the second version of the DSB benchmark – the version used in this challenge – we introduced parallel execution of the benchmark queries, which simulates a real-world workload for the tested system. Also, it is possible to increase/decrease the speed of query issuing, i.e. to modify the speed of the social network activity simulation. More details about this benchmark can be found in [4] and [7].

Task 3: Versioning RDF Data Benchmark. The aim of this task was to test the ability of versioning systems to efficiently manage evolving Linked Data datasets and queries evaluated across multiple versions of these datasets. To do so, we developed the <u>S</u>emantic <u>P</u>ublishing <u>V</u>ersioning <u>B</u>enchmark (SPVB). SPVB acts like a Benchmark Generator, as it generates both the data and the queries needed to test the performance of the versioning systems. It is not tailored to any versioning strategy (the way that versions are stored) and can produce data of different sizes, that can be altered in order to create arbitrary numbers of versions

---

[2] http://www.ldbcouncil.org/.

using configurable insertion and deletion ratios. The data generator of SPVB
uses the data generator of Linked Data Benchmark Council (LDBC) Semantic
Publishing Benchmark[3] (SPB) as well as real DBpedia[4] data. The generated
SPARQL queries are of different types (eight in total) and are partially based on
a subset of the 25 query templates defined in the context of DBpedia SPARQL
Benchmark[5] (DBPSB).

Task 4: Faceted Browsing Benchmark. This goal of this task is to determine the performance and correctness characteristics of triple stores in faceted
browsing scenarios. Faceted browsing thereby stands for a session-based (state-
dependent) interactive method for query formulation over a multi-dimensional
information space. For this purpose, we developed the *Hobbit Faceted Browsing
Benchmark* which analyses SPARQL-based navigation through a dataset. The
benchmark's data generator supports generation of datasets in the transporta-
tion domain. While the schema is fixed, the amount of instance data can be
scaled up to virtually arbitrary sizes. A benchmark run consists of running a
sequence of *scenarios*, whereas a scenario is comprised of a sequence of SPARQL
that simulate related faceted browsing *interactions*, such as adding constraints
or navigating to related set of resources. Conceptually, each interaction causes
a *transition* from the current state of the faceted browsing system to a new one
and thus implies changes in the solution space, i.e. the set of matching resources,
facets, facet values and respective counts. The benchmark defines 11 scenarios
with between 8 and 11 interactions, amounting to 172 queries in total. Every
interaction of a scenario is thereby associated with one out of the 14 *choke-points*
depicted in Fig. 1. A choke-point thus represents a certain type of transition and
is described in more detail in [6]. The overall scores computed by the benchmark are based on the performance and correctness characteristics at the level
of choke-points.

3   Benchmarking Platform

All four tasks are carried out using the HOBBIT benchmarking platform[5]. This
platform offers the execution of benchmarks to evaluate the performance of systems. For every task of the MOCHA challenge, a benchmark was implemented.
The benchmarks are sharing a common API which eases the work of the challenge participants. For the benchmarking of the participant systems a server
cluster has been used. Each of these systems could use up to three servers of
this cluster each of them having 256 GB RAM and 32 cores. This enabled the
benchmarking of monolythical as well as distributed solutions.

---

[3] http://ldbcouncil.org/developer/spb.

[4] http://wiki.dbpedia.org/.

[5] HOBBIT project webpage: http://project-hobbit.eu/,
HOBBIT benchmarking platform: http://master.project-hobbit.eu,
HOBBIT platform source code: https://github.com/hobbit-project/platform.

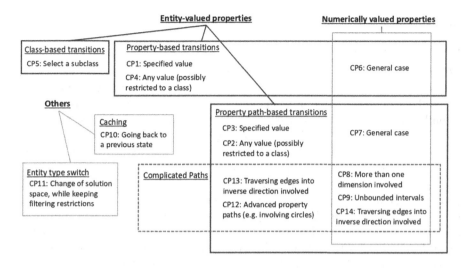

Fig. 1. Overview of the choke-points defined in the faceted browsing benchmark

4   The Challenge

4.1   Overview

The MOCHA2018 challenge ran on 18th May, 2018 and its results were presented during the ESWC 2018 closing ceremony. Five system participated in all tasks:

- Virtuoso v8.0, developed by OpenLink Software.[6] It is a modern enterprise-grade solution for data access, integration, and relational database management.
- Baseline Virtuoso Open Source, developed also by OpenLink Software and is the open source version of Virtuoso.[7]
- Blazegraph[8] that is an ultra-scalable, high-performance graph database with support for the Blueprints and RDF/SPARQL APIs.
- Graph DB Free 8.5 is a family of highly-efficient, robust and scalable RDF databases. It streamlines the load and use of linked data cloud datasets GraphDB Free implements the RDF4J.[9]
- Apache Jena Fuseki 3.6.0 is a SPARQL server. It provides the SPARQL 1.1 protocols for query and update as well as the SPARQL Graph Store protocol.[10]

We had two additional systems participating for task 3, OSTRICH developed by IDLab, Department of Electronics and Information Systems, Ghent University imec and R43ples [3].

---

[6] https://virtuoso.openlinksw.com/.
[7] https://github.com/openlink/virtuoso-opensource.
[8] https://www.blazegraph.com/.
[9] http://rdf4j.org/about/.
[10] https://jena.apache.org/documentation/fuseki2/.

4.2   Results & Discussion

Task 1: Sensor Streams Benchmark

KPIs ODIN's KPIs are divided into two categories: Correctness and Efficiency. Correctness is measured by calculating Recall, Precision and F-Measure. First, the INSERT queries created by each data generator will be send into a triple store by bulk load. After a stream of INSERT queries is performed against the triple store, a SELECT query will be conducted by the corresponding data generator. In Information Retrieval, Recall and Precision were used as relevance measurements and were defined in terms of retrieved results and relevant results for a single query. For our set of experiments, the relevant results for each SELECT query were created prior to the system benchmarking by inserting and querying an instance of the Jena TDB storage solution. Additionally, we will computed Macro and Micro Average Precision, Recall and F-measure to measure the overall performance of the system. Efficiency is measured by using the following metrics: (1) Triples-Per-Second that measures the triples per second as a fraction of the total number of triples that were inserted during a stream. This is divided by the total time needed for those triples to be inserted (begin point of SELECT query - begin point of the first INSERT query of the stream). We provided the maximum value of the triples per second of the whole benchmark. The maximum triples per second value was calculated as the triples per second value of the last stream with Recall value equal to 1. (2) Average Delay of Tasks as the task delay between the time stamp that the SELECT query (task) was sent to the system and the time stamp that the results were send to HOBBIT's storage for evaluation. We report both the average delay of each task and the average delay of task for the whole experiment.

Experiment set-up. For the MOCHA2018 experiments, all parameters were set to their default values, except from the following:

- **Population of generated data** $= 10,000$
- **Number of data generators - agents** $= 4$
- **Number of insert queries per stream** $= 20$

Results for Task 1. Figure 2 illustrates the correctness KPIs for our baseline systems and Virtuoso Commercial 8.0 under the MOCHA2018 configuration. Beginning with the system that had the worse overall performance, Blazegraph, we observe that all KPIs, apart from Micro-Average-Precision, receive the lowest values compared to the other systems. The high value obtained in Micro-Average-Precision indicates that Blazegraph is able to retrieve correct results for the set of SELECT queries that it answered, but its low Macro-Average-Precision value shows that the set of SELECT queries that it managed to retrieve results for was exceptionally low. To continue with, we observe a very similar behavior between Apache Jena Fuseki 3.6.0 and GraphDB Free 8.5. Both systems achieve a high performance in Micro and Macro-Average-Recall, showing that they are able to retrieve all expected results in most SELECT queries. In contrast, their low values in Micro and Macro-Average-Precision indicate that both systems tend to include large sets of irrelevant answers to the query. Finally the two remaining systems, OpenLink Virtuoso and Virtuoso Commercial 8.0 share a similar behavior among the results: they both achieve high Mirco and Macro-Average-Recall and Precision, which shows their superior ability to ingest and retrieve triples with high accuracy.

Regarding their maximum Triples-Per-Second KPI, Fig. 1b shows that Blazegraph achieves the highest value among the other systems. This observation shows that Blazegraph is able to retrieve correct results for a large amount of triples that was inserted in a very short period of time. However, based on Fig. 2, since both Micro and Macro-Average-Recall values are low, we can assume that this situation does not occur very often while using Blazegraph as a triple storage solution.

Regarding the efficiency KPIs, Fig. 1a shows that both OpenLink Virtuoso and Virtuoso Commercial 8.0 require on average a minute amount of time to process the SELECT queries and retrieve correct results. For Apache Jena Fuseki 3.6.0 and GraphDB Free 8.5, the ability to retrieve high Recall performance comes at the cost of efficiency, since both systems have quite high average delay of tasks. Finally, Blazegraph has a low response time compared to the last two systems, but not insignificant as OpenLink Virtuoso and Virtuoso Commercial 8.0.

To conclude, both OpenLink Virtuoso and Virtuoso Commercial 8.0 had received the same Macro-Average-F-Measure value (approx. 0.86), so in order to announce the winner for Task 1 of MOCHA2018, we consider the second KPI in order: Macro-Average-Recall. OpenLink Virtuoso and Virtuoso Commercial 8.0 received 0.88 and 0.92 resp. This clearly indicates that Virtuoso Commercial 8.0 is the winner of Task 1 of MOCHA2018.

Task 2: Data Storage Benchmark

KPIs The key performance indicators for the Data Storage benchmark are rather simple and cover both efficiency and correctness:

- **Bulk Loading Time:** The total time in milliseconds needed for the initial bulk loading of the dataset.
- **Average Task Execution Time:** The average SPARQL query execution time in milliseconds.
- **Average Task Execution Time Per Query Type:** The average SPARQL query execution time per query type in milliseconds.
- **Query failures:** The number of SPARQL SELECT queries whose result set is different (in any aspect) from the result set obtained from the triple store used as a gold standard.
- **Throughput:** The average number of tasks (queries) executed per second.

Experiment set-up. The benchmark had a defined maximum time for the experiment of 3 hours. The DSB parameters used in the challenge were the following:

- **Number of operations** = 15000
- **Scale factor** = 30

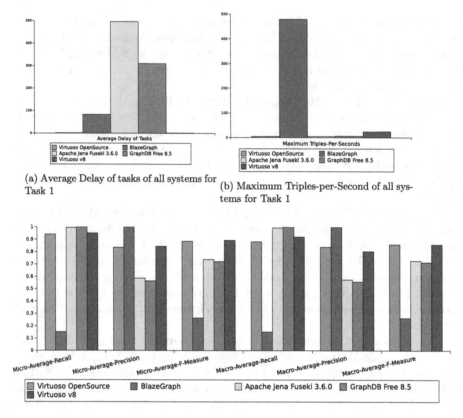

(a) Average Delay of tasks of all systems for Task 1

(b) Maximum Triples-per-Second of all systems for Task 1

Fig. 2. Micro-Average-Recall, Micro-Average-Precision, Micro-Average-F-Measure, Macro-Average-Recall, Macro-Average-Precision, Macro-Average-F-Measure of all systems for Task 1

- **Seed** = 100
- **Time compression ratio (TCR)** = 0.5
- **Sequential Tasks** = false
- **Warm-up Percent** = 20

The scale factor parameter defines the size of dataset. Its value of 30 means 1.4 billion triple dataset. After bulk loading it, there were 15000 SPARQL queries (INSERT and SELECT) executed against the system under test. One fifth of them was used for warm-up, while the rest of the queries were evaluated. The value 0.5 of TCR parameter implies about 17 queries per second.

Results for Task 2. Unfortunately, out of the five systems which participated in the task, only two managed to complete the experiment in the requested time. Blazegraph, GraphDB and Jena Fuseki exhibited a timeout during the bulk loading phase, while the achieved KPIs for Virtuoso v8.0 and Virtuoso Open Source (VOS) are given below.

Based on the results from the main KPIs (Fig. 3), the winning system for the task was Virtuoso 8.0 Commercial Edition by OpenLink Software. In the domain of efficiency, it is 32% faster than VOS regarding average query execution times, and also 6% faster in data loading. In the domain of correctness, Virtuoso v8.0 made 17 query failures compared to the 4 made by VOS; however, having in mind the fact that VOS was used as a golden standard for calculating the expected query results, this KPI is biased, and should not be considered as a weakness of the commercial edition of Virtuoso.

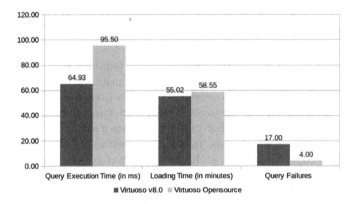

Fig. 3. Task 2: Main KPIs

The rest of the KPIs (Fig. 4) show where the most important advantage comes from. For the complex query types, Virtuoso v8.0 is much faster than its open source counterpart, while the differences in the short look-ups and updates (SPARQL INSERT queries) are negligible.

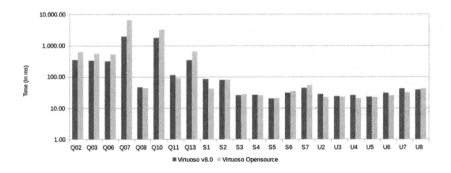

Fig. 4. Task 2: Average Query Execution Time per Query Type

Task 3: Versioning RDF Data Benchmark

KPIs SPVB evaluates the correctness and performance of the system under test through the following *Key Performance Indicators (KPIs)*:

- **Query failures**: The number of queries that failed to execute. Failure refers to the fact that the system under test return a result set (RS_{sys}) that is not equal to the expected one (RS_{exp}). This means that i) RS_{sys} has equal size to RS_{exp} and ii) every row in RS_{sys} has one matching row in RS_{exp}, and vice versa (a row is only matched once). If the size of the result set is larger than 50.000 rows, for time saving, only condition i) is checked.
- **Initial version ingestion speed** (triples/second): the total triples of the initial version that can be loaded per second. We distinguish this from the ingestion speed of the other versions because the loading of the initial version greatly differs in relation to the loading of the following ones, where underlying processes such as, computing deltas, reconstructing versions, storing duplicate information between versions etc., may take place.
- **Applied changes speed** (changes/second): tries to quantify the overhead of such underlying processes that take place when a set of changes is applied to a previous version. To do so, this KPI measures the average number of changes that could be stored by the benchmarked systems per second after the loading of all new versions.
- **Storage space cost** (MB): This KPI measures the total storage space required to store all versions measured in MB.
- **Average Query Execution Time** (ms): The average execution time, in milliseconds for each one of the eight versioning query types.
- **Throughput** (queries/second): The execution rate per second for all queries.

Experiment set-up. Since SPVB gives the ability to the systems under test to retrieve the data of each version as Independent Copies (IC), Change-Sets (CS) or both as IC and CS, three sub-tasks defined in the context of the challenge which only differed in terms of the "generated data form" configuration parameter. So, each participant was able to submit his/her system in the correct sub-task according to the implemented versioning strategy. In particular all the systems benchmarked using the following common parameters:

- **A seed for data generation**: 100
- **Initial version size** (in triples): 200000
- **Number of versions**: 5
- **Version Deletion Ratio** (%): 10
- **Version Insertion Ratio** (%): 15

The experiment timeout was set to 1 hour for all systems.

Results for Task 3. All the systems except of the R43ples one, were managed to be tested. The latest, did not manage to load the data and answer all the queries in the defined timeout of 1 hour.

In order to be able to decide who is the winner for the Task 3 of the challenge, we combined the results of the four most important KPIs and calculated a *final score* that ranges from 0 to 1. Note here that due to reliability issues mentioned earlier, the "Storage space cost" KPI excluded from the final score formula. So, to compute the final score, we assigned weights to those KPIs, whose sum equals to 1. The four KPIs (in order of importance) along with the assigned weights are shown in Table 1.

Table 1. Weights for the four most important KPIs

Order	KPI	Weight
1	Throughput	0.4
2	Queries failed	0.3
3	Initial version ingestion speed	0.15
4	Applied changes speed	0.15

Next, we applied feature scaling[11] to normalize the results of the *Throughput*, *Initial version ingestion speed* and *Applied changes speed* by using the following formula:

$x' = \frac{x - min(x)}{max(x) - min(x)}$, where x is the original value and x' is the normalized value.

Regarding the *Queries failed* KPI, since the lower is the result, the better the system performs, the aforementioned formula applied on the percentage of succeeded queries and not to the number of queries that failed to be executed.

Having all the results normalized in the range of $[0-1]$ and the weights of each KPI, we computed the final scores as the sum of the weighted normalized results. As shown in Fig. 5 VIRTUOSO V8.0 was the system that performed better.

Task 4: Faceted Browsing Benchmark

KPIs The benchmark tracks the conventional indicators for correctness and efficiency, namely precision, recall, F-Measure and query-per-second rate. These measurements are recorded for the individual choke-points as well as the overall benchmark run. For ranking the systems, we were only interested in the latter:

- **Correctness** The conformance of SPARQL query result sets with pre-computed reference results.
- **Performance** The average number of faceted browsing queries per second.

Experiment set-up. The dataset and SPARQL queries generated by the faceted browsing benchmark for the challenge are fully determined by the following settings:

- **Random Seed** = 111
- **Time range** = 0 - 977616000000
- **Number of stops / routes / connections** = 3000 / 3000 / 230000
- **Route length range** = 10 - 50

---

[11] Bring all results into the range of $[0,1]$.

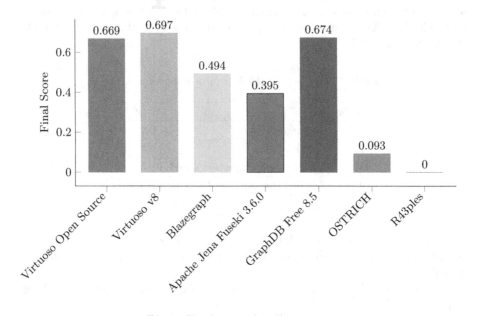

Fig. 5. Final scores for all systems

- **Region size / Cell Density** = 2000 x 2000 / 200
- **Delay Change / Route Choice Power** = 0.02 / 1.3

Results for Task 4. Figure 6 shows the performance results of the faceted browsing benchmark. All participating systems were capable of successfully loading the data and executing the queries. Error values are obtained as follows: For COUNT-based queries, it is the difference in the counts of the actual and reference result. For all other (SELECT) queries, the size of the symmetric difference of the RDF terms mentioned in the reference and actual result sets is taken. While no differences were observer during testing the benchmark with Virtuoso Open Source and Apache Jena Fuseki 3.6.0, surprisingly, very minor ones were observed in the challenge runs. Yet, as all systems nonetheless achieved an f-measure score of >99% we did not discriminate by correctness and ranked them by performance only, depicted in Fig. 6. The winner of the faceted browsing challenge is Virtuoso Open Source with an average rate of 2.3 query executions per second.

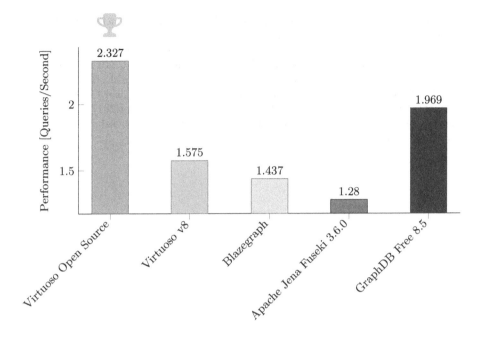

Fig. 6. Task 4 Faceted Browsing Benchmark Results

5 Conclusion

The goal of MOCHA2018 was to test the performance of storage solutions by measuring the systems performance in four different aspects. We benchmarked and evaluated six triple stores and presented a detailed overview and analysis of our experimental set-up, KPIs and results. Overall, our results suggest that the clear winner of MOCHA2018 is Virtuoso v8.0. As part of our future work, we will benchmark more triple storage solutions by scaling over the volume and velocity of the RDF data and use a diverse number of datasets to test the scalability of our approaches.

References

1. Auer, S., Lehmann, J., Ngonga Ngomo, A.-C.: Introduction to linked data and its lifecycle on the web. In: Polleres, A., d'Amato, C., Arenas, M., Handschuh, S., Kroner, P., Ossowski, S., Patel-Schneider, P. (eds.) Reasoning Web 2011. LNCS, vol. 6848, pp. 1–75. Springer, Heidelberg (2011). https://doi.org/10.1007/978-3-642-23032-5_1
2. Erling, O., Averbuch, A., Larriba-Pey, J., Chafi, H., Gubichev, A., Prat, A., Pham, M.-D., Boncz, P.: The LDBC social network benchmark: interactive workload. In: Proceedings of the 2015 ACM SIGMOD International Conference on Management of Data, pp. 619–630. ACM (2015)
3. Graube, M., Hensel, S., Urbas, L.: R43ples: revisions for triples. In: LDQ (2014)

4. Jovanovik, M., Spasić, M.: First version of the data storage benchmark. Project HOBBIT Deliverable 5.1.1 (May 2017)
5. Morsey, M., Lehmann, J., Auer, S., Ngomo, A.-C.N.: DBpedia SPARQL benchmark-performance assessment with real queries on real data. In: ISWC 2011 (2011)
6. Petzka, H., Stadler, C., Katsimpras, G., Haarmann, B., Lehmann, J.: Benchmarking faceted browsing capabilities of triplestores. In: 13th International Conference on Semantic Systems (SEMANTiCS 2017), Amsterdam, 11–14 Sept 2017
7. Spasić, M., Jovanovik, M.: Second version of the data storage benchmark. Project HOBBIT Deliverable 5.1.2 (May 2018)

Versioned Querying with OSTRICH and Comunica in MOCHA 2018

Ruben Taelman[✉], Miel Vander Sande, and Ruben Verborgh

IDLab, Department of Electronics and Information Systems,
Ghent University – IMEC, Gent, Belgium
ruben.taelman@ugent.be

Abstract. In order to exploit the value of historical information in Linked Datasets, we need to be able to store and query different versions of such datasets efficiently. The 2018 edition of the Mighty Storage Challenge (MOCHA) is organized to discover the efficiency of such Linked Data stores and to detect their bottlenecks. One task in this challenge focuses on the storage and querying of versioned datasets, in which we participated by combining the OSTRICH triple store and the Comunica SPARQL engine. In this article, we briefly introduce our system for the versioning task of this challenge. We present the evaluation results that show that our system achieves fast query times for the supported queries, but not all queries are supported by Comunica at the time of writing. These results of this challenge will serve as a guideline for further improvements to our system.

1 Introduction

The Semantic Web [1] of Linked Data [2] is continuously growing and changing over time [3]. While in some cases only the latest version of datasets are required, there is a growing need for access to prior dataset versions for data analysis. For example, analyzing the evolution of taxonomies, or tracking the evolution of diseases in biomedical datasets.

Several approaches have already been proposed to store and query versioned Linked Datasets. However, surveys [4, 5] have shown that there is a need for improved versioning capabilities in the current systems. Existing solutions either perform well for versioned query evaluation, or require less storage space, but not both. Furthermore, no existing solution performs well for all versioned query types, namely querying at, between, and for different versions.

In recent work, we introduced a compressed RDF archive indexing technique [6]— implemented under the name of OSTRICH— that enables highly efficient triple pattern-based versioned querying capabilities. It offers a new trade-off compared to other approaches, as it calculates and stores additional information at ingestion time in order to reduce query evaluation time. This additional information includes pointers to

© Springer Nature Switzerland AG 2018
D. Buscaldi et al. (Eds.): SemWebEval 2018, CCIS 927, pp. 17–23, 2018.
https://doi.org/10.1007/978-3-030-00072-1_2

relevant positions to improve the efficiency of result offsets. Furthermore, it supports efficient result cardinality estimation, streaming results and offset support to enable efficient usage within query engines.

The Mighty Storage Challenge (MOCHA) (https://project-hobbit.eu/challenges/mighty-storage-challenge2018/) is a yearly challenge that aims to measure and detect bottlenecks in RDF triple stores. One of the tasks in this challenge concerns the storage and querying of versioned datasets. This task uses the SPBv [7] benchmark that consists of a dataset and SPARQL query workload generator for different versioned query types. All MOCHA tasks are to be evaluated on the HOBBIT benchmarking platform (https://project-hobbit.eu/). SPBv evaluates SPARQL queries [8], hence we combine OSTRICH, a versioned triple index with triple pattern interface, with Comunica [9], a modular SPARQL engine platform.

The remainder of this paper is structured as follows. First, the next section briefly introduces the OSTRICH store and the Comunica SPARQL engine. After that, we present our preliminary results in Sect. 3. Finally, we conclude and discuss future work in Sect. 4.

2 Versioned Query Engine

In this section we introduce the versioned query engine that consists of the OSTRICH store and the Comunica framework. We discuss these two parts separately in the following sections.

2.1 OSTRICH

OSTRICH is the implementation of a compressed RDF archive indexing technique [6] that offers efficient triple pattern queries in, between, and over different versions. In order to achieve efficient querying for these different query types, OSTRICH uses a hybrid storage technique that is a combination of individual copies, change-based and timestamp-based storage. The initial dataset version is stored as a fully materialized and immutable snapshot. This snapshot is stored using HDT [10], which is a highly compressed, binary RDF representation. All other versions are deltas, i.e., lists of triples that need to be removed and lists of triples that need to be added. These deltas are relative to the initial version, but merged in a timestamp-based manner to reduce redundancies between each version. In order to offer optimization opportunities to query engines that use this store, OSTRICH offers efficient cardinality estimation, streaming results and efficient offset support.

2.2 Comunica

Comunica [9] is a highly modular Web-based SPARQL query engine platform. Its modularity enables federated querying over heterogeneous interfaces, such as SPARQL endpoints [11], Triple Pattern Fragments (TPF) entrypoints [12] and plain RDF files. New types of interfaces and datasources can be supported by implementing an additional software component and plugging it into a publish-subscribe-based system through an external semantic configuration file.

In order to support versioned SPARQL querying over an OSTRICH backend, we implemented (https://github.com/rdfostrich/comunica-actor-rdf-resolve-quad-pattern-ostrich) a module for resolving triple patterns with a versioning context against an OSTRICH dataset. Furthermore, as versions within the SPBv benchmark are represented as named graphs, we rewrite these queries in a separate module (https://github.com/rdfostrich/comunica-actor-query-operation-contextify-version) to OSTRICH-compatible queries in, between, or over different versions as a pre-processing step. Finally, we provide a default Comunica configuration and script (https://github.com/rdfostrich/comunica-actor-init-sparql-ostrich) to use these modules together with the existing Comunica modules as a SPARQL engine. These three modules will be explained in more detail hereafter.

OSTRICH Module. OSTRICH enables versioned triple pattern queries at, between, and for different versions. These query types are respectively known as Version Materialization (VM), Delta Materialization (DM) and Version Querying (VQ). In the context the SPBv benchmark, only the first two query types (VM and DM) are evaluated, which is why only support for these two are implemented in the OSTRICH module at the time of writing.

The OSTRICH Comunica module consists of an actor that enables VM and DM triple pattern queries against a given OSTRICH store, and is registered to Comunica's `rdf-resolve-quad-pattern` bus. This actor will receive messages consisting of a triple pattern and a context. This actor expects the context to either contain VM or DM information, and a reference to an OSTRICH store. For VM queries, a version identifier must be provided in the context. For DM queries, a start and end version identifier is expected.

The `rdf-resolve-quad-pattern` bus expects two types of output:

1. A stream with matching triples.
2. An estimated count of the number of matching triples.

As OSTRICH enables streaming triple pattern query results and corresponding cardinality estimation for all query types, these two outputs can be trivially provided using the JavaScript bindings for OSTRICH (https://github.com/rdfostrich/ostrichnode).

Versioned Query Rewriter Module. The SPBv benchmark represents versions as named graphs. Listings 1 and 2 respectively show examples of VM and DM queries in this representation. Our second module is responsible for rewriting such named-graph-based queries into context-based queries that the OSTRICH module can accept.

```
SELECT ?s ?p ?o WHERE {
  GRAPH <http://graph.version.4> { ?s ?p ?o }
}
```

Listing 1: Version Materialization query for the ?s ?p ?o pattern in version http://graph.version.4 in SPV's named graph representation.

```
SELECT * WHERE {
  GRAPH <http://graph.version.4> { ?s ?p ?o } .
  FILTER (NOT EXISTS {
    GRAPH <http://graph.version.1> { ?s ?p ?o }
  })
}
```

Listing 2: Delta Materialization query for the ?s ?p ?o pattern to get all additions between   version   http://graph.version.1   and   http://graph.version.4 in SPV's named graph representation.

In order to transform VM named-graph-based queries, we detect GRAPH clauses, and consider them to be identifiers for the VM version. Our rewriter unwraps the pattern(s) inside this GRAPHclause, and attaches a VM version context with the detected version identifier.

For transforming DM named-graph-based queries, GRAPH clauses with corresponding FILTER-NOT EXISTS-GRAPH clauses for the same pattern in the same scope are detected. The rewriter unwraps the equal pattern(s), and constructs a DM version context with a starting and ending version identifier. The starting version is always the smallest of the two graph URIs, and the ending version is the largest, assuming lexicographical sorting. If the graph URI from the first pattern is larger than the second graph URI, then the DM queries only additions. In the other case, only deletions will be queried.

SPARQL Engine. The Comunica platform allows SPARQL engines to be created based on a semantic configuration file. By default, Comunica has a large collection of modules to create a default SPARQL engine. For this work, we adapted the default configuration file where we added our OSTRICH and rewriter modules. This allows complete versioned SPARQL querying, instead of only versioned triple pattern querying, as supported by OSTRICH. This engine is available on the npm package manager (https://www.npmjs.com/package/@comunica/actor-init-sparql-ostrich) for direct usage.

3   Evaluation

In this section, we introduce the results of running the SPBv benchmark on Comunica and OSTRICH.

As the MOCHA challenge requires running a system within the Docker-based HOBBIT platform, we provide a system adapter with a Docker container (https://github.com/rdfostrich/challenge-mocha-2018) for our engine that is based on Comunica and OSTRICH. Using this adapter, we ran the SPBv benchmark on our system on the HOBBIT platform with the parameters from Table 1.

Table 1 Configuration of the SPBv benchmark for our experiment.

Parameter	Value
Seed	100
Data form	Changesets
Triples in version 1	100,000
Versions	5
Version deletion ratio	10%
Version addition ratio	15%

For the used configuration, our system is able to ingest 29,719 triples per second for the initial version, and 5,858 per second for the following changesets. The complete dataset requires 17 MB to be stored using our system. The initial version ingestion is significantly faster because the initial version is stored directly as a HDT snapshot. For each following changeset, OSTRICH requires more processing time as it calculates and stores additional metadata and converts the changeset to one that is relative to the initial version instead of the preceding version.

For the 99 queries that were evaluated, our system failed for 27 of them according to the benchmark. The majority of failures is caused by incomplete SPARQL expression support in Comunica, which is not on par with SPARQL 1.1 at the time of writing. The other failures (in task 5.1) are caused by an error in the benchmark where changes in literal datatypes are not being detected. We are in contact with the benchmark developer to resolve this.

For the successful queries, our system achieves fast query evaluation times for all query types, as shown in Table 2. In summary, the query of type 1 (queries starting with a 1-prefix) completely materializes the latest version, type 2 queries within the latest version, type 3 retrieves a full past version, type 4 queries within a past version, type 5 queries the differences between two versions, and type 8 queries over two different versions. Additional details on the query types can be found in the SPBv [7] article.

Table 2 Evaluation times in milliseconds and the number of results for all SPBv queries that were evaluated successfully.

Query	Time	Results	Query	Time	Results
1.1	34,071	141,782	4.5	197	708
2.1	49	128	4.6	1,119	25
2.2	59	32	5.1	13,871	59,229
2.3	27	12	8.1	59	171
2.5	233	969	8.2	56	52
2.6	1,018	19	8.3	31	22
3.1	18,591	100,006	8.4	44	0
2.6	230	46	8.5	709	2,288
4.1	37	91	8.6	8,258	346
4.2	43	16			
4.3	21	2			

4 Conclusions

This article represents an entry for the versioning task in the Mighty Storage Challenge 2018 as part of the ESWC 2018 Challenges Track. Our work consists of a versioned query engine with the OSTRICH versioned triple store and the Comunica SPARQL engine platform. Preliminary results show fast query evaluation times for the queries that are supported. The list of unsupported queries is being used as a guideline for the further development of OSTRICH and Comunica.

During the usage of the SPBv benchmark, we identified several KPIs that are explicitly supported by OSTRICH, but were not being evaluated at the time of writing. We list them here as a suggestion to the benchmark authors for future work:

- Measuring storage size after each version ingestion.
- Reporting of the ingestion time of each version separately, next of the current average.
- Evaluation of querying all versions at the same time, and retrieving their applicable versions.
- Evaluation of stream-based query results and offsets, for example using a dieffi-ciency metric [13].

In future work, we intend to evaluate our system using different configurations of the SPBv benchmark, such as increasing the number of versions and increasing the change ratios. Furthermore, we intend to compare our system with other similar engines, both at triple index-level, and at SPARQL-level.

References

1. Berners-Lee, T., Hendler, J., Lassila, O.: The semantic web. Sci. Am. **284**, 28–37 (2001)
2. Bizer, C., Heath, T., Berners-Lee, T.: Linked data the story so far. In: Semantic Services, Interoperability and Web Applications: Emerging Concepts, pp. 205–227 (2009)
3. Umbrich, J., Decker, S., Hausenblas, M., Polleres, A., Hogan, A.: Towards dataset dynamics: change frequency of linked open data sources. In: 3rd International Workshop on Linked Data on the Web (LDOW) (2010)
4. Fernández, J.D., Polleres, A., Umbrich, J.: Towards efficient archiving of dynamic linked open data. In: Debattista, J., d'Aquin, M., Lange, C. (eds.) Proceedings of the First DIACHRON Workshop on Managing the Evolution and Preservation of the Data Web, pp. 34–49 (2015)
5. Papakonstantinou, V., Flouris, G., Fundulaki, I., Stefanidis, K., Roussakis, G.: Versioning for linked data: archiving systems and benchmarks. In: BLINK ISWC (2016)
6. Taelman, R., Vander Sande, M., Verborgh, R.: OSTRICH: versioned random-access triple store. In: Proceedings of the 27th International Conference Companion on World Wide Web (2018)
7. Papakonstantinou, V., Flouris, G., Fundulaki, I., Stefanidis, K., Roussakis, G.: SPBv: benchmarking linked data archiving systems. In: Proceedings of the 2nd International Workshop on Benchmarking Linked Data and NLIWoD3: Natural Language Interfaces for the Web of Data (2017)
8. Harris, S., Seaborne, A., Prud'hommeaux, E.: SPARQL 1.1 Query Language. W3C, http://www.w3.org/TR/2013/REC-sparql11-query-20130321/ (2013)
9. Taelman, R., Van Herwegen, J., Vander Sande, M., Verborgh, R.: Comunica: a modular SPARQL query engine for the web. In: Proceedings of the 17th International Semantic Web Conference (2018)
10. Fernández, J.D., Martínez-Prieto, M.A., Gutiérrez, C., Polleres, A., Arias, M.: Binary RDF representation for publication and exchange (HDT). Web Semant.: Sci. Serv. Agents World Wide Web **19**, 22–41 (2013)
11. Feigenbaum, L., Todd Williams, G., Grant Clark, K., Torres, E.: SPARQL 1.1 Protocol. W3C, Protocol. W3C, http://www.w3.org/TR/2013/REC-sparql11-protocol-20130321/ (2013)
12. Verborgh, R., et al.: Triple pattern fragments: a low-cost knowledge graph interface for the web. J. Web Semant. **37–38**, 184–206 (2016)
13. Acosta, M., Vidal, M.-E., Sure-Vetter, Y.: Diefficiency metrics: measuring the continuous efficiency of query processing approaches. In: d'Amato, C., Fernandez, M., Tamma, V., Lecue, F., Cudré-Mauroux, P., Sequeda, J., Lange, C., Heflin, J. (eds.) ISWC 2017. LNCS, vol. 10588, pp. 3–19. Springer, Cham (2017). https://doi.org/10.1007/978-3-319-68204-4_1

Benchmarking Virtuoso 8 at the Mighty Storage Challenge 2018: Challenge Results

Milos Jovanovik[1,2] and Mirko Spasić[1,3(✉)]

[1] OpenLink Software, London, United Kingdom
{mjovanovik,mspasic}@openlinksw.com
[2] Faculty of Computer Science and Engineering, Ss. Cyril and Methodius University
in Skopje, Skopje, Macedonia
[3] Faculty of Mathematics, University of Belgrade, Belgrade, Serbia

Abstract. Following the success of Virtuoso at last year's Mighty Storage Challenge - MOCHA 2017, we decided to participate once again and test the latest Virtuoso version against the new tasks which comprise the MOCHA 2018 challenge. The aim of the challenge is to test the performance of solutions for SPARQL processing in aspects relevant for modern applications: ingesting data, answering queries on large datasets and serving as backend for applications driven by Linked Data. The challenge tests the systems against data derived from real applications and with realistic loads, with an emphasis on dealing with changing data in the form of streams or updates. Virtuoso, by OpenLink Software, is a modern enterprise-grade solution for data access, integration, and relational database management, which provides a scalable RDF Quad Store. In this paper, we present the final challenge results from MOCHA 2018 for Virtuoso v8.0, compared to the other participating systems. Based on these results, Virtuoso v8.0 was declared as the overall winner of MOCHA 2018.

Keywords: Virtuoso · Mighty storage challenge
MOCHA · Benchmarks · Data storage · Linked data · RDF · SPARQL

1 Introduction

Last year's Mighty Storage Challenge, MOCHA 2017, was quite successful for our team and Virtuoso – we won the overall challenge [6, 11]. Building on that, we decided to participate in this year's challenge as well, in all four challenge tasks: (i) RDF data ingestion, (ii) data storage, (iii) versioning and (iv) browsing. The Mighty Storage Challenge 2018[1] aims to provide objective measures for how well current systems perform on real tasks of industrial relevance, and also help detect bottlenecks of existing systems to further their development towards practical

---

[1] https://project-hobbit.eu/challenges/mighty-storage-challenge2018/.

© Springer Nature Switzerland AG 2018
D. Buscaldi et al. (Eds.): SemWebEval 2018, CCIS 927, pp. 24–35, 2018.
https://doi.org/10.1007/978-3-030-00072-1_3

usage. This arises from the need for devising systems that achieve acceptable performance on real datasets and real loads, as a subject of central importance for the practical applicability of Semantic Web technologies.

2 Virtuoso Universal Server

Virtuoso Universal Server[2] is a modern enterprise-grade solution for data access, integration, and relational database management. It is a database engine hybrid that combines the functionality of a traditional relational database management system (RDBMS), object-relational database (ORDBMS), virtual database, RDF, XML, free-text, web application server and file server functionality in a single system. It operates with SQL tables and/or RDF based property/predicate graphs. Virtuoso was initially developed as a row-wise transaction oriented RDBMS with SQL federation, i.e. as a multi-protocol server providing ODBC and JDBC access to relational data stored either within Virtuoso itself or any combination of external relational databases. Besides catering to SQL clients, Virtuoso has a built-in HTTP server providing a DAV repository, SOAP and WS* protocol end-points and dynamic web pages in a variety of scripting languages. It was subsequently re-targeted as an RDF graph store with built-in SPARQL and inference [2,3]. Recently, the product has been revised to take advantage of column-wise compressed storage and vectored execution [1].

The largest Virtuoso applications are in the RDF and Linked Data domains, where terabytes of RDF triples are in use – a size which does not fit into main memory. The space efficiency of column-wise compression was the biggest incentive for the column store transition of Virtuoso [1]. This transition also made Virtuoso a competitive option for relational analytics. Combining a schemaless data model with analytics performance is an attractive feature for data integration in scenarios with high schema volatility. Virtuoso has a shared cluster capability for scaling-out, an approach mostly used for large RDF deployments.

A more detailed description of Virtuoso's triple storage, the compression implementation and the translation of SPARQL queries into SQL queries, is available in our paper from MOCHA 2017 [11].

3 Evaluation

Here we present the official challenge results for Virtuoso v8.0 for all MOCHA 2018 tasks, based on the challenge data and benchmark parameters specified by the organizers. Additionally, we give a brief comparison of Virtuoso v8.0 with the other participating systems.

The results presented in this section are publicly available at the official HOB-BIT platform[3], where the challenge and all its tasks took place. The platform allows execution of different benchmarks in order to evaluate the performance of

---

[2] https://virtuoso.openlinksw.com/.

[3] https://master.project-hobbit.eu/.

different systems. A specific benchmark was implemented and used for each task of the MOCHA challenge. These benchmarks share a common API which eases the work of the challenge participants. For the benchmarking of the participant systems, a server cluster was used. Each of the systems could use up to three servers of the cluster, each of them having 256 GB RAM and 32 CPU cores. This enabled benchmarking of both monolithic and distributed solutions.

The Virtuoso v8.0 configuration parameters which we used for the challenge are available at GitHub[4].

Compared to MOCHA 2017, the tasks of MOCHA 2018 were significantly more demanding and the datasets were larger, which lead to a tougher playground. But, this tougher playground is also a better representation of the real-world applications over large amounts of RDF and Linked Data, which the benchmarks and the challenge aim to test.

3.1   Task 1 - RDF Data Ingestion

The aim of this task is to measure the performance of SPARQL query processing systems when faced with streams of data from industrial machinery in terms of efficiency and completeness. This benchmark, called ODIN (StOrage and Data Insertion beNchmark), increases the size and velocity of RDF data used, in order to evaluate how well a system can store streaming RDF data obtained from the industry. The data is generated from one or multiple resources in parallel and is inserted using SPARQL INSERT queries. At some points in time, SPARQL SELECT queries check the triples that are actually inserted and test the system's ingestion performance and storage abilities [4,5].

Table 1. ODIN Configuration

Parameter	Value
Duration of the benchmark	600000
Name of mimicking algorithm	TRANSPORT_DATA
Name of mimicking algorithm output folder	output_data/
Number of data generators - agents	4
Number of insert queries per stream	20
Number of task generators - agents	1
Population of generated data	10000
Seed for mimicking algorithm	100

Results: Our system, Virtuoso v8.0 Commercial Edition, was tested against ODIN as part of the MOCHA challenge. The task organizers specified the bench-

---

[4] https://github.com/hobbit-project/DataStorageBenchmark/blob/master/system/virtuoso.ini.template.

mark parameters for the challenge and their values are shown in Table 1, while the achieved KPIs for our system are presented in the Table 2.

The task has three KPIs:

- **Triples per Second**: For each stream, a fraction of the total number of triples that were inserted during that stream divided by the total time needed for those triples to be inserted.
- **Average Answer Time**: A delay between the time that the SELECT query has been generated and sent to the System Adapter and the time that the results are received by the Evaluation Storage.
- **Correctness**: A recall, precision and F-measure of each SELECT query by comparing the retrieved results and the expected ones obtained from an instance of the Jena TDB storage solution.

Table 2. ODIN KPIs for Virtuoso v8.0

KPI	Value
Average Delay of Tasks (in seconds)	1.3398
Average Triples-Per-Second	11.1909
Macro-Average-F-Measure	0.8587
Macro-Average-Precision	0.8047
Macro-Average-Recall	0.9206
Maximum Triples-Per-Second	25.0410
Micro-Average-F-Measure	0.8945
Micro-Average-Precision	0.8442
Micro-Average-Recall	0.9511

We can divide the KPIs into two categories: Efficiency (Average Delay of Tasks, Average Triples-Per-Second and Maximum Triples-Per-Second) and Correctness (Macro/Micro Average F-Measure/Precision/Recall). In terms of efficiency, the Average Triples-Per-Second KPI is not relevant, as it mainly depends from the benchmark parameters, and the values achieved by all tested systems are very similar. The Average Delay of Tasks KPI shows the efficiency of the system executing SPARQL SELECT queries, i.e. the average execution time of SELECT queries, while the Maximum Triples-Per-Second KPI indicates how fast can tripes be received by the system without any loss.

Our system takes about a second to process a SELECT query on average. Virtuoso Opensource (VOS) is very similar, while all other participating systems show worse results by two orders of magnitude (83-496s). Our achieved value for Maximum Triples-Per-Second is 25, while most of the systems are around 5. Blazegraph achieves the highest value here, but its average recall and f-measure values are exceptionally low. Figures 1 and 2 showcase these results and identify the overall winner of the task. Our system and VOS achieve very similar results

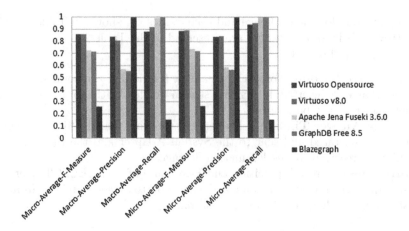

Fig. 1. Task 1 - Correctness KPIs

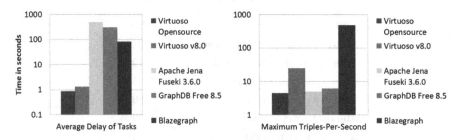

Fig. 2. Task 1 - Efficiency KPIs

in terms of correctness, while Virtuoso v8.0 is undoubtedly better in terms of efficiency. The organizers confirmed this by declaring our system as the winner in this task.

3.2 Task 2 - Data Storage

This task uses the Data Storage Benchmark (DSB) and its goal is to measure how data storage solutions perform with interactive read SPARQL queries, accompanied with a high insert data rate via SPARQL UPDATE queries. This approach mimics realistic application scenarios where read and write operations are bundled together. It also tests systems for their bulk load capabilities [7,12].

Results: The benchmark parameters for the task are shown in Table 3, and the achieved KPIs for our system are presented in Table 4.

The main KPIs of the task are:

– **Bulk Loading Time**: The total time in milliseconds needed for the initial bulk loading of the dataset.

Table 3. DSB Configuration

Parameter	Value
Scale Factor	30
Time Compression Ratio	0.5
Warm-up Percent	20
Enable/Disable Query Type	01100111011010
Number of Operations	15000
Enable Sequential Tasks	False
Seed	100

- **Average Query Execution Times per Query Type**: The execution time is measured for every single query, and for each query type the average query execution time is calculated.
- **Query Failures**: The number of returned results that are not as expected, obtained from the triple store used as a gold standard.

Table 4. DSB KPIs for Virtuoso v8.0

KPI	Value	KPI	Value
Average Query Execution Time	64.9332	Average S5 Execution Time	20.0040
Average Q02 Execution Time	344.6176	Average S6 Execution Time	30.6064
Average Q03 Execution Time	329.1667	Average S7 Execution Time	43.5408
Average Q06 Execution Time	314.3750	Average Update2 Execution Time	27.6351
Average Q07 Execution Time	1943.5962	Average Update3 Execution Time	23.7248
Average Q08 Execution Time	45.9784	Average Update4 Execution Time	25.5625
Average Q10 Execution Time	1761.2794	Average Update5 Execution Time	22.6334
Average Q11 Execution Time	112.3680	Average Update6 Execution Time	30.2792
Average Q13 Execution Time	336.1515	Average Update7 Execution Time	41.9760
Average S1 Execution Time	84.2704	Average Update8 Execution Time	38.2564
Average S2 Execution Time	78.9440	Loading Time (in ms)	3301449
Average S3 Execution Time	25.5088	Query Failures	17
Average S4 Execution Time	26.0392	Throughput (queries/s)	17.6797

In this task, the organizers wanted to stress the scalability of the system by specifying a large dataset with over 1.4 billion triples. Virtuoso justified its dominance with a huge victory in this task, showing why it was well known to be a scalable system. Unfortunately, Blazegraph, GraphDB, and Jena were not able to even load the dataset in the maximum experiment time of 3 hours, thus the only comparable system here was VOS. The comparison of these two systems is given in the Figs. 3, 4 and 5. In the domain of efficiency, Virtuoso v8.0

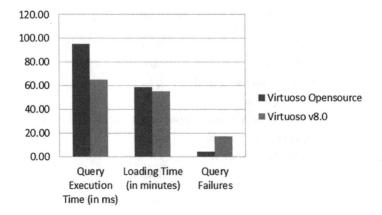

Fig. 3. Task 2 - Main KPIs

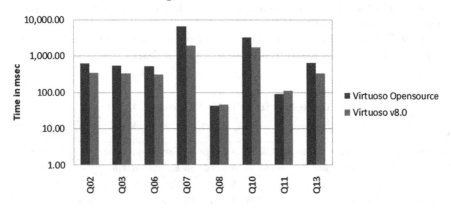

Fig. 4. Task 2 - Average complex query execution time per type

is 32% faster than VOS regarding average query execution times, and around 6% faster in data loading. In the domain of correctness, Virtuoso v8.0 made 17 query failures compared to the 4 made by VOS; however, having in mind the fact that VOS was used as a golden standard for calculating the expected query results, this KPI is biased, and should not be considered as a weakness of the latest version of Virtuoso.

3.3   Task 3 - Versioning RDF Data

The aim of this task is to test the ability of versioning systems to efficiently manage evolving datasets, where triples are added or deleted, and queries evaluated across the multiple versions of said datasets. It uses the Versioning Benchmark (VB) [8,9].

Results: Table 5 shows the benchmark configuration and Table 6 shows the results achieved by Virtuoso v8.0 for the versioning task. The evaluation is based on the following performance KPIs:

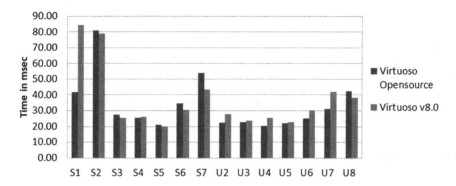

Fig. 5. Task 2 - Average short and update query execution time per type

Table 5. Versioning benchmark configuration

Parameter	Value
Generated Data Form	IC/CS/IC+CS
Initial Version Size (in triples)	200000
Number of Versions	5
Version Deletion Ratio (%)	10
Version Insertion Ratio (%)	15
A seed for data generation (%)	100

– **Query Failures**: The number of queries whose returned results are not those that were expected.
– **Throughput** (in queries per second): The execution rate per second for all queries.
– **Initial Version Ingestion Speed** (in triples per second): The total triples that can be loaded per second for the dataset's initial version.
– **Applied Changes Speed** (in triples per second): The average number of changes that can be stored by the benchmarked system per second after the loading of all new versions.
– **Average Query Execution Time** (in ms): The average execution time, in milliseconds, for each one of the eight versioning query types.

Apart from the same five systems which participated in the previous tasks, two additional ones (specialized in storing different versions of datasets) took part at Task 3. These two system, as well as some other versioning systems are not mature enough to handle very large datasets, so the organizers decided to decrease the dataset size and give them the opportunity to compete. Relatively small datasets result in comparable throughput achieved by our and the other systems (Fig. 6). Judging by the previous tasks, we would expect for our system to have better throughput compared the rest of the participants if the datasets were larger. Still, Virtuoso v8.0 acted better in the initial loading, applied

Table 6. Versioning benchmark KPIs for virtuoso v8.0

KPI	Value	KPI	Value
Applied Changes (changes/s)	19831.2070	QT5, Avg. Exec. Time (ms)	7186.0000
Initial Ingestion (triples/s)	43793.0820	QT6, Avg. Exec. Time (ms)	165.7500
QT1, Avg. Exec. Time (ms)	16861.0000	QT7, Avg. Exec. Time (ms)	17.0000
QT2, Avg. Exec. Time (ms)	147.6667	QT8, Avg. Exec. Time (ms)	155.6389
QT3, Avg. Exec. Time (ms)	11944.0000	Queries Failed	2
QT4, Avg. Exec. Time (ms)	142.3889	Throughput (queries/s)	1.2068

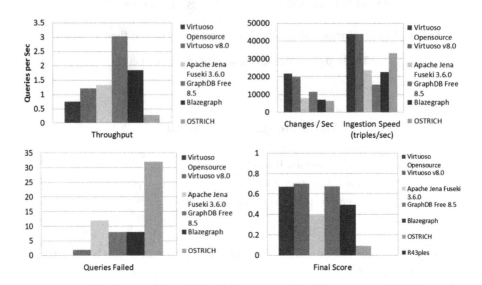

Fig. 6. Task 3 - KPIs and final score

changes per second and correctness (Fig. 6). Therefore, the announcement of the winner here was not straightforward, and the organizers combined the results of the four most important KPIs with their assigned weights (Throughput - 0.4, Queries Failed - 0.3, Initial Version Ingestion Speed - 0.15 and Applied Changes Speed - 0.15). With this, they calculated the final scores which range from 0 to 1. The final scores for all participant systems are shown on Fig. 6, with Virtuoso v8.0 providing the best performance in the task.

3.4   Task 4 - Faceted Browsing

This task uses the Faceted Browsing Benchmark, which tests existing solutions for their capabilities of enabling faceted browsing through large-scale RDF datasets, that is, it analyses their efficiency in navigating through large datasets, where the navigation is driven by intelligent iterative restrictions. The goal of the task is to measure the performance relative to dataset characteristics, such as overall size and graph characteristics [10].

Table 7. Faceted browsing configuration

Parameter	Value	Parameter	Value
Cells per LatLon	200	Number of Stops	3000
Delay Chance	0.2	Preconfiguration	mocha2018
Final Trip Time	977616000000	Quick Test Run	false
Initial Trip Time	0	Random Seed	111
Maximum Route Length	50	Region Size X	2000
Minimum Route Length	10	Region Size Y	2000
Number of Connections	230000	Route Choice Power	1.3
Number of Routes	3000		

Results: The benchmark parameters for the task are given in Table 7, while Table 8 shows the Virtuoso v8.0 results. The evaluation is based on the following performance KPIs:

- **Correctness**: The conformance of SPARQL query result sets with precomputed reference results in terms of precision, recall and F-Measure.
- **Performance**: Query-per-second rate.

Table 8. Faceted browsing main KPIs for virtuoso v8.0

KPI	Value
Total F-measure	1.0
Total precision	1.0
Total recall	1.0
Throughput (queries/s)	1.5755

The comparison of our system with the other participant systems for the task are shown on the Fig. 7. As we can see, Virtuoso v8.0 did not achieve the best performance score for the task, and the task winner was VOS. One possible reason why our system showed lower performance compared to VOS and GraphDB should be sought in the small size of the dataset and the fact that our system was pre-configured for significantly larger deployments. Even though the training phase of this task used a smaller dataset, we still expected a larger dataset for the challenge run, thus configured Virtuoso v8.0 accordingly.

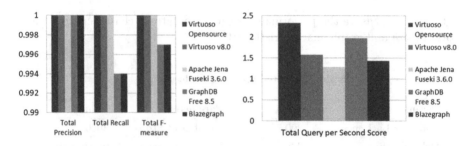

Fig. 7. Task 4 - Correctness (left) and Performance (right)

4   Conclusion and Future Work

In this paper, we provide an overview of the setup, participation and results of
Virtuoso v8.0 at the Mighty Storage Challenge - MOCHA 2018, at the Extended
Semantic Web Conference - ESWC 2018. Our system participated in all four
tasks of the challenge: (i) RDF data ingestion, (ii) data storage, (iii) versioning and (iv) faceted browsing, winning the first three. Therefore, the challenge
organizers declared Virtuoso v8.0 as a clear winner of MOCHA 2018.

As future work, a further Virtuoso evaluation has been planned, using other
dataset sizes and especially larger datasets, stressing its scalability. We can
already foresee improvements of the query optimizer, driven by the current evaluation. The comparison of our performance with the other participant systems
also provides significant guidelines for the future development of Virtuoso.

Acknowledgments. This work has been supported by the H2020 project HOBBIT
(GA no. 688227).

References

1. Erling, O.: Virtuoso, a hybrid RDBMS/graph column store. IEEE Data Eng. Bull.
   35(1), 3–8 (2012)
2. Erling, O., Mikhailov, I.: RDF support in the virtuoso DBMS. Networked
   Knowledge-Networked Media, pp. 7–24. Springer, Berlin (2009)
3. Erling, O., Mikhailov, I.: Virtuoso: RDF support in a native RDBMS. Semantic
   Web Information Management, pp. 501–519. Springer, Berlin (2010)
4. Georgala, K.: First version of the data extraction benchmark for sensor data, May
   2017. Project HOBBIT Deliverable 3.1.1
5. Georgala, K.: Second version of the data extraction benchmark for sensor data,
   May 2018. Project HOBBIT Deliverable 3.1.2
6. Georgala, K., Spasić, M., Jovanovik, M., Petzka, H., Röder, M., Ngomo, A.-C.N.:
   MOCHA2017: the mighty storage challenge at ESWC 2017. In: Semantic Web
   Challenges: 4th SemWebEval Challenge at ESWC 2017, pp. 3–15. Springer, Berlin
   (2017)
7. Jovanovik, M., Spasić, M.: First version of the data storage benchmark, May 2017.
   Project HOBBIT Deliverable 5.1.1

8. Papakonstantinou, V., Fundulaki, I., Flouris, G.: Second version of the versioning benchmark, May 2018. Project HOBBIT Deliverable 5.2.2
9. Papakonstantinou, V., Fundulaki, I., Roussakis, G., Flouris, G., Stefanidis, K.: First version of the versioning benchmark, May 2017. Project HOBBIT Deliverable 5.2.1
10. Petzka, H.: First version of the faceted browsing benchmark, May 2017. Project HOBBIT Deliverable 6.2.1
11. Spasić, M., Jovanovik, M.: MOCHA 2017 as a challenge for virtuoso. In: Dragoni, M., Solanki, M., Blomqvist, E. (eds.) SemWebEval 2017. CCIS, vol. 769, pp. 21–32. Springer, Cham (2017). https://doi.org/10.1007/978-3-319-69146-6_3
12. Mirko Spasić and Milos Jovanovik. Second Version of the Data Storage Benchmark, May 2018. Project HOBBIT Deliverable 5.1.2

Open Knowledge Extraction Challenge

Open Knowledge Extraction Challenge 2018

René Speck[1,2](✉), Michael Röder[2,3], Felix Conrads[3], Hyndavi Rebba[4],
Catherine Camilla Romiyo[4], Gurudevi Salakki[4], Rutuja Suryawanshi[4],
Danish Ahmed[4], Nikit Srivastava[4], Mohit Mahajan[4], and
Axel-Cyrille Ngonga Ngomo[2,3]

[1] Leipzig University, DICE Group, Augustusplatz 10, Leipzig 04109, Germany
speck@informatik.uni-leipzig.de
[2] Institute for Applied Informatics (InfAI) e.V., AKSW Group, Hainstraße 11,
Leipzig 04109, Germany
{michael.roeder,axel.ngonga}@uni-paderborn.de
[3] Paderborn University, DICE Group, Warburger Straße 100,
Paderborn 33098, Germany
conrads@informatik.uni-leipzig.de
[4] Paderborn University, Warburger Straße 100, Paderborn 33098, Germany
{hyndavi,rcathy,gss,rutuja,dahmed,nikit,mac24079}@mail.uni-paderborn.de

Abstract. The fourth edition of the Open Knowledge Extraction Challenge took place at the 15th Extended Semantic Web Conference in 2018. The aim of the challenge was to bring together researchers and practitioners from academia as well as industry to compete of pushing further the state of the art in knowledge extraction from text for the Semantic Web. This year, the challenge reused two tasks from the former challenge and defined two new tasks. Thus, the challenge consisted of tasks such as Named Entity Identification, Named Entity Disambiguation and Linking as well as Relation Extraction. To ensure an objective evaluation of the performance of participating systems, the challenge ran on a version the FAIR benchmarking platform Gerbil integrated in the HOBBIT platform. The performance was measured on manually curated gold standard datasets with Precision, Recall, F1-measure and the runtime of participating systems.

Keywords: Knowledge extraction · Named entity identification ·
Named entity linking · Relation extraction · Semantic web

1 Introduction

The vision of the Semantic Web (SW) is an extension of the Document Web with the goal to allow intelligent agents to access, process, share and understand the data in the web. These agents are build upon structured data. Thus, implementing the vision of the SW requires transforming unstructured and semi-structured

© Springer Nature Switzerland AG 2018
D. Buscaldi et al. (Eds.): SemWebEval 2018, CCIS 927, pp. 39–51, 2018.
https://doi.org/10.1007/978-3-030-00072-1_4

data from the Document Web into structured machine processable data of the SW using knowledge extraction approaches.

To push the state of the art in knowledge extraction from natural language text, the Open knowledge extraction Challenge (OKE) aims to trigger attention from the knowledge extraction community and foster their broader integration with the SW community. Therefore, the OKE has the ambition to provide a reference framework for research on knowledge extraction from text for the SW by defining a number of tasks (typically from information and knowledge extraction), taking into account specific SW requirements.

The first OKE 2015 [7] and second OKE 2016 [8] were both composed of two tasks, *Entity Recognition, Linking and Typing for Knowledge Base population* and *Class Induction and entity typing for Vocabulary and Knowledge Base enrichment*. In the first version, the challenge had four participants, Adel [9], CETUS [12], FRED [2] and OAK@Sheffield [4]. In the second version the challenge had five participants, a new version of Adel [10], Mannheim [3], WestLab-Task1 [1], WestLab-Task2 [5] and the baseline with CETUS from the former year. In the third version, the OKE 2017 [14] was composed of three tasks, *Focused Named Entity Identification and Linking, Broader Named Entity Identification and Linking* and *Focused Musical Named Entity Recognition and Linking*. In this version the challenge had two participants, a new version of Adel [11] and the baseline with FOX [13].

This year, the OKE 2018 reused the first two tasks from the former challenge, *Focused* and *Broader Named Entity Identification and Linking* as well as defined two new tasks, *Relation Extraction* and *Knowledge Extraction*.

The rest of this paper is structured as follows: We begin with defining preliminaries in Sect. 2 before describing the challenge tasks in Sect. 3. In Sect. 4 we give a brief introduction of the participating systems and compare the results achieved by our evaluation on the gold datasets in Sect. 5. Finally, we discuss the insights provided by the challenge and possible extensions in the last section.

2   Preliminaries and Notations

In this section we define terminologies and notations that are used throughout this paper.

Knowledge Base

Let a knowledge base K consists of a set of entities E_K, an entity type hierarchy T_K with a function that maps each entity to its types $\psi_K : E_K \rightarrow 2^{T_K}$, a relation type hierarchy R_K with a function that maps each relation to its domain and range entity types $\phi_K : R_K \rightarrow T_K \times T_K$ and relation instances or facts $F_K = \{r(e_1, e_2)\} \subset R_K \times E_K \times E_K$ with $r \in R_K$ and $(e_1, e_2) \in E_K \times E_K$.

Named Entity Identification

Consider each dataset D to be a set of documents and each document d to be sequence of words $d = (w_j)_{j=1,2,...}$ The identification of named entities in a given document d aims to find named entity mentions $M = \{m_i\}_{i=1,2,...}$ that express named entities. A named entity mention m is a sequence of words in d identified by its start and end index $I_M = \{(a, b)_i\}_{i=1}^{|M|}$ where $a, b \in \mathbb{N}$ and $a < b$.

Named Entity Disambiguation and Linking

The aim of named entity disambiguation and linking to a knowledge base K is to assign each named entity mention $m \in M$ to an entity in K if possible, otherwise to generate a new resource for such an emerging entity, i.e. $\varphi : M \to E_K \cup E_{\bar{K}}$ is a function that maps an entity mention to an entity in E_K or to a newly generated entity in $E_{\bar{K}}$ for an emerging entity that does not exist in K.

Closed Binary Relation Extraction

Closed binary relation extraction aims to find relations $r(e_j, e_k)$ expressed in a given text $d \in D$ with $r \in R_K$ and $e_j, e_k \in E_K \cup E_{\bar{K}}$. Often, closed binary relation extraction is limited to a subset of relations $R \subset R_K$ in K.

RDF/Turtle Prefixes

Listing 1.1 depicts the RDF/Turtle prefixes for all in- and output examples we illustrate in this paper.

```
@prefix rdf: <http://www.w3.org/1999/02/22-rdf-syntax-ns#> .
@prefix xsd: <http://www.w3.org/2001/XMLSchema#> .
@prefix itsrdf: <http://www.w3.org/2005/11/its/rdf#> .
@prefix nif: <http://persistence.uni-leipzig.org/nlp2rdf/ontologies/nif-core#> .
@prefix rdfs: <http://www.w3.org/2000/01/rdf-schema#> .
@prefix dbr: <http://dbpedia.org/resource/> .
@prefix dbo: <http://dbpedia.org/ontology/> .
@prefix aksw: <http://aksw.org/notInWiki/> .
@prefix oa: <http://www.w3.org/ns/oa#> .
```

Listing 1.1: Prefixes for the examples.

3   Open Knowledge Extraction Challenge Tasks

In this section, we describe each of the four challenge tasks and provide examples for a better understanding. All tasks depended on the DBpedia knowledge base. A participating system was not expected to process any preprocessing steps (e.g. pronoun resolution) on the input data. In case a resource for an entity was missing in the knowledge base, a system was expected to generate a URI using a namespace that does not match a known knowledge base (e.g. http://aksw.org/notInWiki/) for this emerging entity.

We carried out the evaluation with the HOBBIT benchmarking platform and the benchmark implementation of the HOBBIT project[1] which relies on the Gerbil evaluation framework [15].

3.1   Task 1: Focused Named Entity Identification and Linking

The first task compromised a two-step process with (a) the identification of named entity mentions in sentences and (b) the disambiguation of these mentions by linking to resources in the given knowledge base. A competing system was expected to (a) identify named entity mentions $\{m_i\}_{i=1,2,...}$ in a given document d with $m_i \in d$ by the start and end indices $\{(a,b)_i\}_{i=1,2,...}$. Further, (b) to find the URIs in K to disambiguate and link each mention if possible. Otherwise, URIs should be generated the emerging entities and link these mentions, $\{\varphi(m_i)\}_{i=1,2,...}$.

This task was limited to a subset T of entity types[2] provided by the DBpedia knowledge base, i.e. $T := \{$dbo:Person, dbo:Place, dbo:Organisation$\}$.

Example Listing 1.2 is an example request document of task 1 and Listing 1.3 is the expected response document for the given request document.

```
<http://www.ontologydesignpatterns.org/data/oke-challenge-2018/task-1/sentence-1#char=0,90>
  a nif:RFC5147String , nif:String , nif:Context ;
  nif:beginIndex "0"^^xsd:nonNegativeInteger ;
  nif:endIndex "90"^^xsd:nonNegativeInteger ;
  nif:isString "Leibniz was born in Leipzig in 1646 and attended the University of Leipzig
      from 1661-1666."@en .
```

Listing 1.2: Example request document in task one.

```
<http://www.ontologydesignpatterns.org/data/oke-challenge-2018/task-1/sentence-1#char=0,7>
  a nif:RFC5147String , nif:String ;
  nif:anchorOf "Leibniz"@en ;
  nif:beginIndex "0"^^xsd:nonNegativeInteger ;
  nif:endIndex "7"^^xsd:nonNegativeInteger ;
  nif:referenceContext <http://www.ontologydesignpatterns.org/data/oke-challenge-2018/task-1/
      sentence-1#char=0,90> ;
  itsrdf:taIdentRef dbr:Gottfried_Wilhelm_Leibniz .

<http://www.ontologydesignpatterns.org/data/oke-challenge-2018/task-1/sentence-1#char=20,27>
  a nif:RFC5147String , nif:String ;
  nif:anchorOf "Leipzig"@en ;
  nif:beginIndex "20"^^xsd:nonNegativeInteger ;
  nif:endIndex "27"^^xsd:nonNegativeInteger ;
  nif:referenceContext <http://www.ontologydesignpatterns.org/data/oke-challenge-2018/task-1/
      sentence-1#char=0,90> ;
  itsrdf:taIdentRef dbr:Leipzig .

<http://www.ontologydesignpatterns.org/data/oke-challenge-2018/task-1/sentence-1#char=53,74>
  a nif:RFC5147String , nif:String ;
  nif:anchorOf "University of Leipzig"@en ;
  nif:beginIndex "53"^^xsd:nonNegativeInteger ;
  nif:endIndex "74"^^xsd:nonNegativeInteger ;
  nif:referenceContext <http://www.ontologydesignpatterns.org/data/oke-challenge-2018/task-1/
      sentence-1#char=0,90> ;
```

---

[1] http://project-hobbit.eu.

[2] http://mappings.dbpedia.org/server/ontology/classes (full type hierarchy).

```
itsrdf:taIdentRef dbr:Leipzig_University .
```

Listing 1.3: Example of the expected response document in task one.

3.2 Task 2: Broader Named Entity Identification and Linking

This task extended the former task towards a broader set of entity types. Beside the three types of the first task, a competing system had to identify other types of entities and to link these entities as well.

In the first column in Table 1, a complete list of types that are considered in this task is provided. The middle column contains example subtypes of the corresponding type if any such type is available and the last column contains example instances in the knowledge base for the types.

Table 1: Types, subtype examples and instance examples. All types are defined in the `dbo` namespace.

Type	Subtypes	Instances
`Activity`	`Game, Sport`	`Baseball,Chess`
`Agent`	`Organisation, Person`	`Leipzig_University`
`Award`	`Decoration, NobelPrize`	`Humanitas_Prize`
`Disease`		`Diabetes_mellitus`
`EthnicGroup`		`Javanese_people`
`Event`	`Competition, PersonalEvent`	`Battle_of_Leipzig`
`Language`	`ProgrammingLanguage`	`English_language`
`MeanOfTransportation`	`Aircraft, Train`	`Airbus_A300`
`PersonFunction`	`PoliticalFunction`	`PoliticalFunction`
`Place`	`Monument, WineRegion`	`Beaujolais, Leipzig`
`Species`	`Animal, Bacteria`	`Cat, Cucumibacter`
`Work`	`Artwork, Film`	`Actrius, Debian`

3.3 Task 3: Relation Extraction

Given a dataset D with documents, the DBpedia knowledge base K, a target entity type hierarchy T with $T \subset T_K$ and a target relation type hierarchy R with $R \subset R_K$. Furthermore, annotations of the documents are given, i.e., entity mentions M with the positions I_M, the disambiguation $\varphi : M \to E_K \cup E_{\bar{K}}$ and the types of entities $\psi_K : E_K \cup E_{\bar{K}} \to T$.

The aim of this task was to find binary relations $r(e_j, e_k)$ with $r \in R$ and $e_j, e_k \in E_K \cup E_{\bar{K}}$. The domain and range entity types for the applied relations in this task, $\phi_K : R \to T \times T$ were given and are depicted in Table 2.

For the preparation of an output document, a participating system had to serialize the found binary relations in the input document with RDF statements using the the Web Annotation Vocabulary to connect the extracted statement with the given document.

Example Listing 1.4 is an example request document of this task and Listing 1.5 shows the triples that have to be added to the request to form the expected response document.

```
<http://www.ontologydesignpatterns.org/data/oke-challenge-2018/task-3/sentence-1#char=0,78>
  a nif:RFC5147String , nif:String , nif:Context ;
  nif:beginIndex "0"^^xsd:nonNegativeInteger ;
  nif:endIndex "78"^^xsd:nonNegativeInteger ;
  nif:isString "Conor McGregor's longtime trainer, John Kavanagh, is ready to shock the world
      ."^^xsd:string .

<http://www.ontologydesignpatterns.org/data/oke-challenge-2018/task-3/sentence-1#char=0,22>
  a nif:RFC5147String , nif:String , nif:Phrase ;
  nif:anchorOf "Conor McGregor's"^^xsd:string ;
  nif:beginIndex "0"^^xsd:nonNegativeInteger ;
  nif:endIndex "22"^^xsd:nonNegativeInteger ;
  nif:referenceContext <http://www.ontologydesignpatterns.org/data/oke-challenge-2018/task-3/
      sentence-1#char=0,78> ;
  its:taClassRef dbo:Person ;
  itsrdf:taIdentRef dbr:Conor_McGregor .

<http://www.ontologydesignpatterns.org/data/oke-challenge-2018/task-3/sentence-1#char=35,48>
  a nif:RFC5147String , nif:String , nif:Phrase ;
  nif:anchorOf "John Kavanagh"^^xsd:string ;
  nif:beginIndex "35"^^xsd:nonNegativeInteger ;
  nif:endIndex "48"^^xsd:nonNegativeInteger ;
  nif:referenceContext <http://www.ontologydesignpatterns.org/data/oke-challenge-2018/task-3/
      sentence-1#char=0,78> ;
  its:taClassRef dbo:Person ;
  itsrdf:taIdentRef aksw:John_Kavanagh .
```

Listing 1.4: Example request document in task three.

```
[]
  a rdf:Statement , oa:Annotation ;
  rdf:object dbr:Conor_McGregor ;
  rdf:predicate dbo:trainer ;
  rdf:subject aksw:John_Kavanagh ;
  oa:hasTarget [
    a oa:SpecificResource;
    oa:hasSource <http://www.ontologydesignpatterns.org/data/oke-challenge-2018/task-3/
    sentence-1#char=0,78> ] .
```

Listing 1.5: Example of the expected response document in task three.

3.4 Task 4: Knowledge Extraction

This task was a combination of task one *Focused Named Entity Identification and Linking* and task three *Relation Extraction.*

In this task, the input documents comprised only natural language text, similar to the input documents of task one. Thus, without annotations of entity mentions, entity types and linkings to the knowledge base. A participating system was expected to provide combined serializations with the information analogous to task one and task three.

Table 2: Relation type hierarchy with domain and range entity types. All relations and types are defined in the dbo namespace.

Relation	Domain	Range
affiliation	Organisation	Organisation
almaMater	Person	EducationalInstitution
bandMember	Band	Person
birthPlace	Person	Place
ceo	Organisation	Person
child	Person	Person
club	Athlete	SportsTeam
country	Organisation, Person, Place	Country
deathPlace	Person	Place
debutTeam	Athlete	SportsTeam
department	PopulatedPlace	PopulatedPlace
district	Place	PopulatedPlace
doctoralAdvisor	Scientist	Person
doctoralStudent	Scientist	Person
employer	Person	Organisation
formerBandMember	Band	Person
formerTeam	Athlete	SportsTeam
foundationPlace	Organisation	City
headquarter	Organisation	PopulatedPlace
hometown	Organisation, Person	Settlement
leaderName	PopulatedPlace	Person
locatedInArea	Place	Place
location	Organisation, Person, Place	Place
nationality	Person	Country
parent	Person	Person
president	Organisation	Person
relative	Person	Person
spouse	Person	Person
subsidiary	Company	Company
tenant	ArchitecturalStructure	Organisation
trainer	Athlete	Person

Example Listing 1.6 is an example request document of this task and Listing 1.7 shows the triples that have to be added to the request to form the expected response document.

```
<http://www.ontologydesignpatterns.org/data/oke-challenge-2018/task-4/sentence-1#char=0,78>
  a nif:RFC5147String , nif:String , nif:Context ;
  nif:beginIndex "0"^^xsd:nonNegativeInteger ;
  nif:endIndex "78"^^xsd:nonNegativeInteger ;
  nif:isString "Conor McGregor's longtime trainer, John Kavanagh, is ready to shock the world
      ."^^xsd:string .
```

Listing 1.6: Example request document in task four.

```
<http://www.ontologydesignpatterns.org/data/oke-challenge-2018/task-4/sentence-1#char=0,22>
  a nif:RFC5147String , nif:String , nif:Phrase ;
  nif:anchorOf "Conor McGregor's"^^xsd:string ;
  nif:beginIndex "0"^^xsd:nonNegativeInteger ;
  nif:endIndex "22"^^xsd:nonNegativeInteger ;
  nif:referenceContext <http://www.ontologydesignpatterns.org/data/oke-challenge-2018/task-4/
      sentence-1#char=0,78> ;
  itsrdf:taIdentRef dbr:Conor_McGregor .

<http://www.ontologydesignpatterns.org/data/oke-challenge-2018/task-4/sentence-1#char=35,48>
  a nif:RFC5147String , nif:String , nif:Phrase ;
  nif:anchorOf "John Kavanagh"^^xsd:string ;
  nif:beginIndex "35"^^xsd:nonNegativeInteger ;
  nif:endIndex "48"^^xsd:nonNegativeInteger ;
  nif:referenceContext <http://www.ontologydesignpatterns.org/data/oke-challenge-2018/task-4/
      sentence-1#char=0,78> ;
  itsrdf:taIdentRef aksw:John_Kavanagh .

[]
  a rdf:Statement , oa:Annotation ;
  rdf:object dbr:Conor_McGregor ;
  rdf:predicate dbo:trainer ;
  rdf:subject aksw:John_Kavanagh ;
  oa:hasTarget [
    a oa:SpecificResource;
    oa:hasSource <http://www.ontologydesignpatterns.org/data/oke-challenge-2018/task-4/
    sentence-1#char=0,78> ] .
```

Listing 1.7: Example of the expected response document in task four.

4 Participants

The challenge attracted five research groups this year. Four groups from universities and one from industry. Two groups participated in the challenge and competed in task three. Both systems, RelExt and the Baseline, are briefly described in the next subsections.

4.1 RelExt

RelExt is an approach based on a deep learning classifier that uses self attention. The classifier was trained on sentences from Wikipedia pages chosen in a distance supervised fashion with the DBpedia knowledge base. RelExt uses a filtering step to find words in sentences that might express specific relations. These words are manually filtered by the authors and were used to refine the sentences to obtain training data.

RelExt participated in task three of the OKE challenge.

4.2 Baseline

The baseline system for task three simply used the annotated documents in the evaluation phase without a learning or training step on the training dataset. The input documents of task three consisted of annotated entities with entity linkings to the DBpedia knowledge base. Thus, the baseline chose pairwise the given URIs of the linked entities from the input documents to create a SPARQL query to request all predicates that hold between two URIs in DBpedia. In case two entities had a statement in the knowledge base with a predicate included in the task, the baseline chose this statement in the response document.

For instance, let "Leibniz was born in Leipzig." be a sentence in a document together with the entity linkings `dbr:Gottfried_Wilhelm_Leibniz` for the entity mention "Leibniz" and `dbr:Leipzig` for the entity mention "Leipzig" in this example sentence. The baseline took both resource URIs and queried the DBpedia to find all predicates that hold between these resources. In this case, the predicate `dbo:birthPlace` is in DBpedia as well as in the tasks list of predicates and thus chosen to be in the response document.

5 Evaluation

In this section we describe the evaluation of the participating systems. We define the evaluation metrics, describe the datasets and present the evaluation results.

5.1 Platform

The benchmarking platform for named entity recognition and linking implemented within HOBBIT [6][3] reuses some of the concepts developed within the open source project Gerbil. These concepts were migrated and adapted to the HOBBIT architecture. The platform calculates the micro and macro averages of Precision, Recall and F1-measure. Additionally, the time a system needs to answer a request as well as the number of failing requests (i.e. requests that are answered with an error code instead of a valid response) are determined.

5.2 Measures

Equations 1, 2 and 3 formalize Precision p_d, Recall r_d and F1-measure used to evaluate the quality of the systems responses for each document $d \in D$. They consist of the number of true positives TP_d, false positives FP_d and false negatives FN_d. We micro and macro average the performances over the documents.[4]

$$p_d = \frac{TP_d}{TP_d + FP_d} \tag{1}$$

---

[3] http://project-hobbit.eu/wp-content/uploads/2017/04/D2.2.1.pdf.
[4] The macro averages for the performance measures can be retrieved from the official HOBBIT SPARQL endpoint at http://db.project-hobbit.eu/sparql.

$$r_d = \frac{TP_d}{TP_d + FN_d} \tag{2}$$

$$f_d = 2 \cdot \frac{p_d \cdot r_d}{p_d + r_d} \tag{3}$$

5.3   Datasets

The datasets for all tasks were manually curated. For the first two tasks, we reused cleansed and updated versions of the datasets from the former year, i.e. fixed typos and wrong indices. For the two new tasks, we created new datasets with eight annotators. Each annotator reviewed the annotations of another annotator to reduce noise and cognitive bias as well as to ensure a high quality.

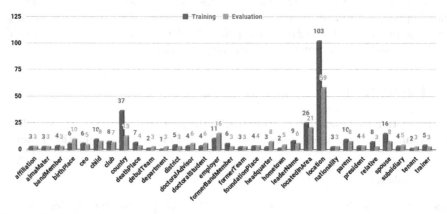

Fig. 1: Predicate distribution of the training and evaluation datasets for challenge task three and four.

In Fig. 1, the predicate distributions of the training and evaluation datasets for the challenge task three and four are shown. The training dataset consisted of 319 and the evaluation dataset of 239 annotated relations in the documents (cf. Table 3). In Table 3 the number of annotated documents, entities, entity linkings and relations are shown for the challenge datasets. The datasets for the first two tasks were without relation annotations and the datasets for the last two tasks were the same but differ as input documents for participating systems, cf. the examples in Listings 1.4 and 1.6. For instance, the training datasets for task one consisted of 60 documents with 377 annotated entity mentions as well as linkings of these mentions to 255 unique resources in the DBpedia knowledge base or newly generated resources for emerging entities.

The training datasets for each task are available at the challenge website.[5]

---

[5] https://project-hobbit.eu/challenges/oke2018-challenge-eswc-2018.

Table 3: Attributes and their values of the challenge datasets.

	Task 1	Task 2	Task 3 and Task 4
	Training/Evaluation	Training/Evaluation	Training/Evaluation
Documents	60/58	56/53	100/100
Entities	377/381	422/462	308/274
Linkings	255/255	301/327	254/242
Relations	N/A	N/A	319/239

5.4 Results

Table 4 depicts the results of the OKE 2018 on task three. The results are available in Hobbit for both participants, RelExt[6] and the Baseline[7]. RelExt won this task with 54.30% Macro F1-Score and 48.01% Micro F1-Score.

Table 4: RelExt and Baseline.

KPI	RelExt	Baseline
Avg. ms per Doc	836.26	513.42
Error count	1	0
Macro F1-Score	**54.30**	8.00
Macro precision	53.98	10.00
Macro recall	64.17	7.18
Micro F1-Score	**48.01**	8.66
Micro precision	39.62	68.75
Micro recall	60.92	4.62

6   Conclusion

The OKE attracted five research groups from academia and industry coming from the knowledge extraction as well as the SW communities. Indeed, the challenge proposal was aimed at attracting groups from these two communities in order to further investigate existing overlaps between both. Additionally, one of the goals of the challenge was to foster the collaboration between the two communities, to the aim of growing further the SW community. To achieve this goal we defined a SW reference evaluation framework, which is composed of (a) four tasks, (b) a training and evaluation dataset for each task, and (c) an evaluation

---

[6] https://master.project-hobbit.eu/experiments/1529075533385.
[7] https://master.project-hobbit.eu/experiments/1527777181515.

framework to measure the performance of the systems. Although the participation in terms of number of competing systems remained quite limited with two, we believe that the challenge is a success in the hybridisation of Semantic Web technologies with knowledge extraction methods.

As a matter of fact, the evaluation framework is available online and can be reused by the community and for next editions of the challenge.

Acknowledgements. This work has been supported by the H2020 project HOBBIT (no. 688227), the BMWI projects GEISER (no. 01MD16014E) and OPAL (no. 19F2028A), the EuroStars projects DIESEL (no. 01QE1512C) and QAMEL (no. 01QE1549C).

References

1. Chabchoub, M., Gagnon, M., Zouaq, A.: Collective disambiguation and semantic annotation for entity linking and typing. In: Sack, H., Dietze, S., Tordai, A., Lange, C. (eds.) SemWebEval 2016. CCIS, vol. 641, pp. 33–47. Springer, Cham (2016). https://doi.org/10.1007/978-3-319-46565-4_3
2. Consoli, S., Recupero, D.R.: Using FRED for named entity resolution, linking and typing for knowledge base population. In: Gandon, F., Cabrio, E., Stankovic, M., Zimmermann, A. (eds.) SemWebEval 2015. CCIS, vol. 548, pp. 40–50. Springer, Cham (2015). https://doi.org/10.1007/978-3-319-25518-7_4
3. Faralli, S., Ponzetto, S.P.: Dws at the 2016 open knowledge extraction challenge: a hearst-like pattern-based approach to hypernym extraction and class induction. In: Sack, H., Dietze, S., Tordai, A., Lange, C. (eds.) Semantic Web Challenges, pp. 48–60. Springer International Publishing, Cham (2016)
4. Gao, J., Mazumdar, S.: Exploiting linked open data to uncover entity types. In: Gandon, F., Cabrio, E., Stankovic, M., Zimmermann, A. (eds.) SemWebEval 2015. CCIS, vol. 548, pp. 51–62. Springer, Cham (2015). https://doi.org/10.1007/978-3-319-25518-7_5
5. Haidar-Ahmad, L., Font, L., Zouaq, A., Gagnon, M.: Entity typing and linking using SPARQL patterns and DBpedia. In: Sack, H., Dietze, S., Tordai, A., Lange, C. (eds.) SemWebEval 2016. CCIS, vol. 641, pp. 61–75. Springer, Cham (2016). https://doi.org/10.1007/978-3-319-46565-4_5
6. Ngomo, A.C.N., Röder, M.: HOBBIT: holistic benchmarking for big linked data. In: ESWC, EU Networking Session (2016). http://svn.aksw.org/papers/2016/ESWC_HOBBIT_EUNetworking/public.pdf
7. Nuzzolese, A.G., Gentile, A.L., Presutti, V., Gangemi, A., Garigliotti, D., Navigli, R.: Open knowledge extraction challenge. In: Gandon, F., Cabrio, E., Stankovic, M., Zimmermann, A. (eds.) SemWebEval 2015. CCIS, vol. 548, pp. 3–15. Springer, Cham (2015). https://doi.org/10.1007/978-3-319-25518-7_1
8. Nuzzolese, A.G., Gentile, A.L., Presutti, V., Gangemi, A., Meusel, R., Paulheim, H.: The second open knowledge extraction challenge. In: Sack, H., Dietze, S., Tordai, A., Lange, C. (eds.) SemWebEval 2016. CCIS, vol. 641, pp. 3–16. Springer, Cham (2016). https://doi.org/10.1007/978-3-319-46565-4_1
9. Plu, J., Rizzo, G., Troncy, R.: A hybrid approach for entity recognition and linking. In: Gandon, F., Cabrio, E., Stankovic, M., Zimmermann, A. (eds.) SemWebEval 2015. CCIS, vol. 548, pp. 28–39. Springer, Cham (2015). https://doi.org/10.1007/978-3-319-25518-7_3

10. Plu, J., Rizzo, G., Troncy, R.: Enhancing entity linking by combining NER models. In: Sack, H., Dietze, S., Tordai, A., Lange, C. (eds.) SemWebEval 2016. CCIS, vol. 641, pp. 17–32. Springer, Cham (2016). https://doi.org/10.1007/978-3-319-46565-4_2

11. Plu, J., Troncy, R., Rizzo, G.: ADEL@OKE 2017: a generic method for indexing knowledge bases for entity linking. In: Dragoni, M., Solanki, M., Blomqvist, E. (eds.) SemWebEval 2017. CCIS, vol. 769, pp. 49–55. Springer, Cham (2017). https://doi.org/10.1007/978-3-319-69146-6_5

12. Röder, M., Usbeck, R., Speck, R., Ngomo, A.C.N.: Cetus - a baseline approach to type extraction. In: 1st Open Knowledge Extraction Challenge @ 12th European Semantic Web Conference (ESWC 2015) (2015). http://svn.aksw.org/papers/2015/ESWC_CETUS_OKE_challenge/public.pdf

13. Speck, R., Ngomo, A.-C.N.: Ensemble learning for named entity recognition. In: Mika, P., Tudorache, T., Bernstein, A., Welty, C., Knoblock, C., Vrandečić, D., Groth, P., Noy, N., Janowicz, K., Goble, C. (eds.) ISWC 2014. LNCS, vol. 8796, pp. 519–534. Springer, Cham (2014). https://doi.org/10.1007/978-3-319-11964-9_33

14. Speck, R., Röder, M., Oramas, S., Espinosa-Anke, L., Ngomo, A.-C.N.: Open knowledge extraction challenge 2017. In: Dragoni, M., Solanki, M., Blomqvist, E. (eds.) SemWebEval 2017. CCIS, vol. 769, pp. 35–48. Springer, Cham (2017). https://doi.org/10.1007/978-3-319-69146-6_4

15. Usbeck, R., et al.: GERBIL - general entity annotation benchmark framework. In: 24th WWW conference (2015). http://svn.aksw.org/papers/2015/WWW_GERBIL/public.pdf

Relation Extraction for Knowledge Base Completion: A Supervised Approach

Héctor Cerezo-Costas[✉] and Manuela Martín-Vicente

Gradiant, Edificio CITEXVI, local 14, Universidade de Vigo, Vigo, Spain
hcerezo@gradiant.org
http://www.gradiant.org

Abstract. This paper outlines our approach to the extraction on pre-defined relations from unstructured data (OKE Challenge 2018: Task 3). Our solution uses a deep learning classifier receiving as input raw sentences and a pair of entities. Over the output of the classifier, expert rules are applied to delete known erroneous relations. For training the system we gathered data by aligning DBPedia relations and Wikipedia pages. This process was mainly automatic, applying some filters to refine the training records by human supervision. The final results show that the combination of a powerful classifier model with expert knowledge have beneficial implications in the final performance of the system.

Keywords: Relation extraction · Classification · Deep learning

1 Introduction

In relation extraction from unstructured text the objective is to identify semantic connections between two entities in a short piece of text (e.g. a sentence or a short paragraph). This sub-problem considers that the entities were previously detected and their positions marked in the text, for example, with a Named Entity Recognizer (NER) [8]. Depending on the number and the types of the relations considered in a scenario, a link between two entities could exist or not. Relations are defined as triples in which an entity takes the role of *subject* (the domain of the relation) and the other is the *object* (their range). In theory, a relation must be inferred from the context surrounding the entities within the text, nevertheless, extra information might be available (e.g. entity links on external open data services such as DBPedia [1]). Thus, in order to tackle relation extraction from a document, a system must take as input the text and the entities detected in it and provide as output the whole set of relations that can be extracted from it. For evaluation, either a non-existent relation or an existent relation with the roles swapped will be marked as an error.

The great difficulty of the problem lies in the inherent nature of natural language. Sometimes the relations are very explicit and easy to obtain, but, more regularly, relations can only be inferred with a deep understanding of the context information, or with the aid of external knowledge. The absence of a

© Springer Nature Switzerland AG 2018
D. Buscaldi et al. (Eds.): SemWebEval 2018, CCIS 927, pp. 52–66, 2018.
https://doi.org/10.1007/978-3-030-00072-1_5

large corpus with ground truth data also complicates the task, as many current state of the art approaches in natural language processing require historic data for learning models in order to solve a problem (e.g. machine translation [23] or text classification [19]).

The aim of this challenge is interesting because automatic relation extraction could help to add new information to linked data services or create new ones. It is also the basis of many question-answering solutions [2] and machine comprehension tasks [21]. In those applications, the use of external resources with human curated knowledge improves the overall accuracy over models that do not use that information (e.g. see, for example, [13]). Traditionally, knowledge databases have been generated in a collaborative way [18] or using semi-automatic systems in which human verification is required at some stage [12]. The coverage of these systems tend to be low, hence hindering their use in real applications.

Our approach to the relation extraction problem uses a twofold strategy. On the one hand we employ a state of the art deep learning classifier to tag the relations existing between a pair of entities. Here, we solve the absence of training data using a weak supervised approach, gathering the data from external public resources and in which humans only take part at the final stages –on the process of filtering the training records. On the other hand, we take all the relations extracted in a document and apply common sense heuristics to filter out some relations that seem illogical. Results show that the combination of the individual view of the classifier and the global view of the expert knowledge obtains the best performance in all the tests.

This document is organized as follows. Section 2 summarizes other research proposals that are relevant for our work. In Sect. 3 we depict our solution for the relation extraction problem. The Sect. 4 outlines how we gather the training corpus and how the final model is generated. The criteria for evaluating the system and the results obtained are depicted in Sect. 5. Section 6 explains the results as well as discusses current shortcomings and future directions of our work.

2 Related Work

Knowledge base completion has recently acquired a lot of interest in the academic community. For example, the authors in [17] use a deep learning model to extract new facts from existing entities in a knowledge base. The input to the system consists of the entities and the relation between them. When the system gives a high score to a triple, it creates new links in the database. The authors showed that by representing entities with word vectors and the usage of a deep learning model they can successfully extract new facts in different knowledge bases. However, although the system can reason about facts that can be expressed as triples, it is unable to process free text. A similar approach was employed in [9] to augment curated commonsense resources (in this case ConceptNet [18]). In this work, a neural net model reasons about a triple giving higher probability to those relations deemed to be true. Common sense knowledge is difficult to gather from raw text so taking human curated knowledge as basis for training a system seems a good idea.

One common limitation of these solutions is that they cannot process entities not known beforehand, because a word vector embedding with the knowledge about an entity must exist.

Although the methodology we use is similar, the problem we try to solve differ as our objective is to extract facts from raw text. Previous approaches to this problem in the literature used bootstrapping from a seed of relations from DBPedia to obtain new patterns that carry semantic information of a relation (e.g. $(Person, marry, Person)$ implies $(Person, spouse, Person)$) [6]. A two-step approach was followed in [3]. First, taking the relations from Freebase[1] as basis, the authors search the patterns that typically connect two entities for a relation. Afterwards, they find new entities that match those patterns in raw text in order to augment the database.

In a more constrained scenario with less relations, a WordNet RDF Graph was generated by the authors in [16]. The data sources are dictionary definitions of entities from which different parts are extracted and a semantic role labeling is applied to each segment. The semantic role embeds a relation between the definition segment and the entity. The training data were manually reviewed by humans but the number of records were very low (only 2000).

When the requirements of data for training is higher, other methods employ weak supervision [22] or distant supervision [11] to overcome the problem of the absence of manually labelled ground truth. Subsequently, those samples are used to train a classifier. In general, these approaches make the assumption that every single match between a triple extracted from a knowledge base and a raw sentence is an expression of that relation. However, that is not typically the case with some relations derived from *places* (e.g. *birthPlace, nationality*), because users tend to perform many actions in those locations but the sentence will not embed the meaning of the relation. The authors in [15] built an undirected graphical model that captures both the task of predicting relations between entities, and the task of predicting which sentences express these relations. During training they do not know in which sentences the relation actually holds. Only the assumption that at least one sentence embeds the relation is needed by the system. In our approach we tried a different strategy. We constrained the match of the entities to their Wikipedia pages and we introduced a manual revision of keywords for filtering out the samples not representative of a relation. This supervision improved the final training corpus without having to manually review every single record.

3   Approach

3.1   Overall Architecture

The simplified architecture of the system is depicted in Fig. 1. We model the relation extraction as a classification problem. The system will receive as input

---

[1] Freebase data dumps https://developers.google.com/freebase/.

a NIF-RDF document[2] containing a text message which is a subset of the original web content, a set of entities and their types –limited to the most generic categories: Person, Organization and Location, and the locations of the entities in the given text. The NIF document also contains the URI of the original web content. The NIF format is a collection of specifications and ontologies that facilitates the interchange of information between Natural Language Processing (NLP) resources and tools. Those documents are received through an AMQ message broker and the documents are processed sequentially. To allow communication with external applications the results generated by our system also use the same format.

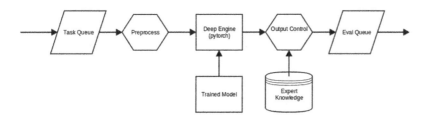

Fig. 1. Architecture of the system.

Although the system consumes and produces NIF-RDF documents, internally the information is preprocessed to be used by the classification engine. The text with the raw sentences is tokenized. Entity names in the query are substituted by a token containing the role and type of the entity, if known (e.g. *"David Beckham married Victoria Adams in 1999"* is substituted to *"Subject_Person married Object_Person in 1999"*). As we cannot know the role of an entity in a relation beforehand, we try the two possible combinations for each pair of entities.

We try to extract all the available information from the NIF document but sometimes some information is missing or some parsing errors occur. If this happens, we try to infer the entity types using the NER capability of Spacy[3]. If the entity types still cannot be extracted, we use the generic Unknown type.

Once the records are sanitized, they are fed to the relation extraction engine. The engine consists of a classifier that extracts relations from the entities in the document and an output control system that use ad-hoc heuristics with expert knowledge to filter those relations that contradict others with higher confidence.

3.2 Classification

The core of the relation extraction engine is a deep learning classifier that uses self-attention [10]. Figure 2 depicts the basic architecture of this neural network. The first layers of the model consist of a standard bidirectional LSTM

---

[2] NLP Interchange Format http://persistence.uni-leipzig.org/nlp2rdf/.

[3] Spacy https://spacy.io/.

network [7]. At each step of the network the model obtains an internal hidden state with the combined representation of previous steps. The attention layers compute a fixed set of vectors as multiple weighted sums of the hidden states from the bidirectional LSTM layers. Finally, the last layers of the network are full dense with the softmax layer as the last step to obtain the labels for each sample.

We used an implementation in *pytorch*[4] that we adapted to add extra capabilities that were not initially present in the system (such as the activation function selection or treating the problem as a query-context problem[5]). We have used the same optimization errors that were used in [10].

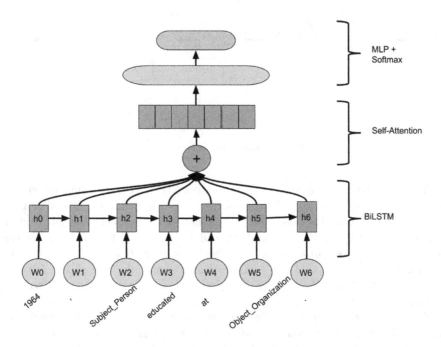

Fig. 2. Classification structure

The possible outputs of the system are the set of predefined relations and the extra *norel* class, which was added to indicate that no relation exists between a pair of entities. Those relations different than *norel* will be passed to the output control module.

3.3 Output Control

Given a global view of the relations linked to an entity in a document, the classifier makes illogical mistakes that can be easily corrected with basic

---

[4] Pytorch http://pytorch.org/.
[5] This last capability was discarded due to its bad performance.

commonsense rules. To this aim, after the classifier model assigns a probability for each possible relation, a heuristic determines the relations that will be finally assigned to a pair of entities. For example, the heuristic knows which relations are allowed for each pair of entities and which are not allowed. Since we substitute real names of entities with a wildcard (consisting of the role and type of the entity) in the preprocessing stage, some relations are difficult to distinguish only from the context (e.g. *Country* vs *Location*). For this reason, we add rules for adding/modifying some relations when a given entity is in a list of countries or when a more generic relation could apply (e.g. adding *location* if *birthPlace* relation is detected). The objective of these heuristics is to improve the output of the system and to overcome some limitations of our approach (e.g. only one relation could be assigned to a given pair of entities).

The basic functionality of the output control module is:

- **Filtering out relations** generated by the deep learning classifier. For example, rules were deemed useful applied to the *dbo:parent/child* relation (e.g. an entity is linked to more than two parents) or *dbo:trainer* (e.g. for a pair of entities a bidirectional relation has been obtained from the classifier). The module also filters duplicates. Duplicates might be obtained when an entity with the same reference, appears in different parts of the raw text, with different anchor[6].
- **Creating new relations** only when they are directly implied by some existing relation already extracted by the classifier. The most straightforward cases are the *dbo:birthPlace*, *dbo:deathPlace*, *dbo:country*, etc. that also imply the *dbo:location* relation.

The complete set of rules used by the output control module are depicted in Table 1. When the system has to discard relations (e.g. when a bidirectional *trainer* relation has been extracted between two entities whereas this should not comply with the heuristics), it filters out those with the lowest probability, given by the output of the classifier. Neither these heuristics are very specific nor they use external resources (except the list of countries) so they could be easily extrapolated to other relations in the future without much effort.

The triples that were not filtered out and those newly cre ated will conform the final output of the system.

4   Model Training

4.1   Data Preparation

One of the biggest challenges we had to confront was that we needed a big corpus to train our deep learning supervised model. Manually creating such corpus is extremely costly in time and human effort, and therefore it is unfeasible in most cases.

We followed a process in several steps with little human intervention to obtain a corpus of considerable size:

---

[6] The "anchor" is the textual representation of an entity in a sentence.

Table 1. Set of heuristics employed to filter the output

Heuristics	Relations that apply
Removing duplicates	All
Not more than x predicates	*Parent*
Not bidirectional	*Trainer*
Directly implied by other relation	*Country, birthPlace, deathPlace* → *location* *ceo* → *employer* (reversed)
Removing relations that does not hold	*Child* if *parent* deleted
Country and object is a country	Directly implied *locatedInArea*

- **Gathering triples from DBPedia.** We have used public DBPedia dumps for obtaining the initial corpus for training. We extracted triples from the file *mappingbased_objects_en.ttl*. Only those triples with relations in the benchmark were filtered for training. From this subset, we extracted the list of unique entities that take part in at least one relation of the corpus.
- **Obtaining data types of entities.** We have used the file *instance_types_en.ttl* to obtain the data type of an entity. Data types are further processed to take the type of highest order following the DBPedia ontology (*Person, Organisation* and *Place*). We have used an extra wildcard type to tag those entities whose type was unknown. Although the more specific types could greatly simplify the task of relation extraction, the benchmarks in which our system was tested employed only generic types of entities in the input.
- **Extracting sentences from Wikipedia pages.** Finally, for each pair of entities linked by a relation, we extracted their respective Wikipedia pages and selected the sentences in which both entities appeared. We allowed partial matching of entities with certain restrictions (e.g. we consider *Beckham* for *David Beckham* as a positive match but we did not allow the match *North* for *North America*). If both entities in a relation share some words (e.g. *David Beckham* and *Victoria Beckham*), we only allow partial matches of the entity that coincides with the Wikipedia page. The match could exist in a sequence of text of more than one sentence, limiting it to a paragraph. Entities in the sentences are substituted by tags representatives of the role of the entity in the relationship (*subject* or *object*). Both entities must appear in the selected text. Otherwise the sample was discarded.

After a careful inspection of the corpus, we detected that some sentences did not carry the meaning of the relationship. That was especially true with relationships indicative of location (e.g. *birthPlace* or *deathPlace*). For example, the subject could study or work in a city that was her birthplace but this sentence did not contain, explicitly or implicitly, any useful information about this topic. Hence, an extra filtering step was executed in the corpus. We set a seed of words representative of each relationship and tried to find iteratively other frequent words (adjectives, verbs and nouns) co-occurring with the seeds that might act as triggers. At the end of each iteration we manually filtered out the words that are erroneously marked as triggers. Triggers might be shared among different relationships (e.g. *bandMember* and *formerBandMember* have many trigger words in common). That was the only point during the training process in which humans had to perform some supervision.

Those trigger words are finally used to refine the original corpus to obtain the final training data. Figure 3 shows the distribution of records of the training data (1261165 records). As it can be clearly seen, relations relative to places (*Country*, *birthPlace* or *location*) dominate the corpus.

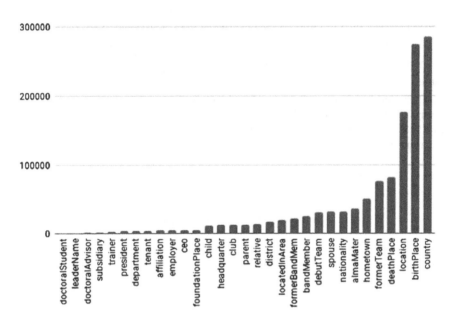

Fig. 3. Distribution of relations in the training corpus.

4.2   Training

Given two entities with their types and a short text in which both entities appear, we model the problem as a multiclass classification in which the objective is to extract the best relationship among a possible set. We consider 32 classes (31 relationships and one extra label that means no relationship - *norel*).

To build negative samples for the *norel* class we use the following strategy. For each training sample we exchange the roles of the entities and mark it as "norel". However there are several cases in which this is not possible. Two of the relationships are bidirectional: *spouse* and *relative*. Furthermore, there are two pairs of relationships comprising opposing relationships (*child-parent* and *doctoralStudent-doctoralAdvisor*). We take that into account when transforming the data for training. Table 2 shows one example of each type.

The most relevant parameters of the model are summarized in Table 3. Due to the highly unbalanced training data, we apply weights to the different classes during training. 100.000 records from the training data are reserved for validation. The algorithm converged after only six epochs. The model was trained in just one GPU Nvidia GeForce GTX TITAN X.

Table 2. Training samples

Description	Label
Subject_Person died in Object_Place in August 2001 at the age of 86	deathPlace
Object_Person died in Subject_Place in August 2001 at the age of 86	Norel
Subject_Person was Object_Person's third wife	Spouse
Object_Person was Subject_Person's third wife	Spouse
In 1943, Subject_Person disowned his daughter Object_Person for marrying the English actor director and producer Charlie Chaplin	Child
In 1943, Object_Person disowned his daughter Subject_Person for marrying the English actor director and producer Charlie Chaplin	Parent

Table 3. Classification Parameters

Parameter	Value	Parameter	Value
Words in dictionary	300.000	Embedding size	300
Hidden units (GRU)	300	Attention units	150
Attention hops	30	Dropout	0.5
Class number	32	Batch size	32
Activation function	Relu	Input size (classif. layer)	1500
Learning rate (Adam)	0.001	Number of layers (GRU)	2

5 Evaluation

The system is tested using the Hobbit Platform[7]. This platform provides a common benchmarking framework to researchers coming from both industry and academia. In this platform systems must be embedded in a Docker container. They must implement a GERBIL compatible API [20] in order to receive the tasks and provide the appropriate answers to the benchmark. Users do not have

---

[7] https://project-hobbit.eu/.

direct access to the results during the evaluation so the framework provides an interesting way for fair comparison among the different systems.

Two different datasets were used for testing the system. The *training dataset* was available during development and the *eval dataset* was released after the evaluation had started. Each dataset is composed of 100 documents in the NIF format. Each document contains at least two entities. Not all pair of entities must produce a relation. All documents in the *training* and *test set* contain at least one valid relation.

5.1   Results

The results of the system in training data given by the hosts of the competition –that we used as validation data rather that for training due to the small number of samples – and evaluation data are depicted in Table 4. As it can be seen the results of the deep learning classifier with output control outperforms clearly the standalone version, when only the classifier decisions are taken into account. Furthermore, the computational overhead of the system with the output control is negligible. The reader might notice that the deep learning classifier was executed in CPU at test time whereas it was trained in GPU because we were not aware that the remote Hobbit platform offered GPU support.

Table 4. Results on training (validation) data and evaluation data. The abbreviatures DL and DL + OC stands for deep learning classifier and deep learning classifier with output control.

KPIs	Training data		Eval data	
	DL	DL + OC	DL	DL + OC
Avg. ms per doc	1170.0	1196.94	818.535	836.26
Error count	0	0	1	1
Macro F1	0.239	0.364	0.492	0.543
Macro precision	0.268	0.356	0.546	0.54
Macro recall	0.254	0.407	0.523	0.642
Micro F1	0.236	0.385	0.421	0.48
Micro precision	0.233	0.321	0.413	0.396
Micro recall	0.238	0.482	0.428	0.609

The number of relations filtered and created by the system with output control can be seen in Table 5. The relations produced by this module are concentrated in two relations *dbo:location* and *dbo:locatedInArea*. These relations are only generated when another less generic relation that implies location exists between the pair of entities (e.g. *dbo:birthPlace*, *dbo:country*, etc.). The total number of relations produced by the DL + OC system are 404 and 342 relations in training and evaluation data respectively. That means the OC system contributes with the 40% of the relations in the training set and the 33% of the relations in the evaluation set. With the *dbo:location* the relations created increased to the 74% and 56%. The trained model typically selects the more specific relation most of the times and the *dbo:location* relation is created most of the times by the output control. Hence, the effect of the output control is not negligible despite its simplicity.

Table 5. Relations deleted and created by the output control in training and evaluation data

Relation	Training data		Eval data	
	Deleted	New	Deleted	New
dbo:child	1	0	0	0
dbo:country	21	8	10	13
dbo:hometown	2	0	3	0
dbo:nationality	4	0	2	0
dbo:parent	1	0	0	0
dbo:spouse	4	2	2	0
dbo:trainer	3	0	1	0
dbo:ceo	0	1	1	0
dbo:employer	0	9	0	12
dbo:locatedInArea	0	46	0	36
dbo:location	0	96	2	53
dbo:affiliation	0	0	1	0
dbo:birthPlace	0	0	2	0
dbo:club	0	0	1	0
Total	36	162	25	114

6   Discussion

The combination of expert knowledge in rules and machine learning classifiers has showed strong performance. It is true that adding these heuristics add extra overhead in terms of human supervision but their generation can be easily automatized due to their simplicity and their replicability among the different relations.

The approach we have used is to a great extent conditioned by the data that could be gathered from external resources. Although using a classifier that can output more than one label might seem preferable for solving this task, gathering training data for this system is unfeasible with current resources. DBPedia relations are far from being complete and thus we cannot be sure that all the relations within a pair of entities already exist in the knowledge base. Thus, if that were chosen strategy we will have obtained many negative samples for some relations due to the incompleteness of the knowledge base.

The absence of a large corpus of ground truth data for training/validating the system is a big handicap in relation extraction. The lack of training records is very common in NLP community, especially as the problems they deal with grow in complexity. That is the case, for example, of abstractive summarization in which the inherent structure of digital news (description, abstract and headlines) is used to train and test the different approaches [14].

DBPedia relations were the best resource we could find for solving this task. However, DBPedia relations are collaboratively generated and they will likely differ from those that will be required during real evaluation with expert generated samples. For example, the *dbo:relative* relation is populated with many cartoon and TV show characters (e.g. Buzz Lightyear and Jessy, the cowgirl). Some of them indeed exhibit some kinship but the samples extracted would not be easily extrapolated to real relatives. Other relations show a sample set that will differ from those used in validation (e.g. the *dbo:trainer* relation samples in DBPedia refer mostly to WWE fighters or boxers but the training and eval data contained samples from other sports: soccer, swimming, tennis that were not very popular among the training samples).

The verb tense also differ among the different samples. All the examples in the evaluation dataset are obtained from blog post and news where the present is used whereas the training data are from encyclopedic resources in which the past tense is more common (in training records the *left* keyword has 3658 appearances whilst the *leave/es/ing* is only in 704 samples).

Our system did not use additional knowledge information with the only exception of a list of countries. In some examples, using external information as features for the classification is the only way to discern the real relation between a pair of entities. For example, the sample *Zayn Malik Leaves One Direction* will translate to *ent_x_Person Leaves ent_x_Organization*. The relations *dbo:formerTeam* and *dbo:formerBandMember* could match perfectly this sentence. Only by knowing that *Zayn Malik* is a singer or *One Direction* a music group the relation can be unequivocally disambiguated.

Using extra information as context could also help taking the decisions. The NIF document contains information that was not used by our system. For example, the NIF contains the URI of the original web content from which the raw sentences were extracted. In order to use the information the link between an entity an their anchors in the context must be generated. These links between anchors and entities are only provided in the text selected. Obtaining the best results, implies the use of coreference resolution techniques [5] (e.g. for pronoun resolution).

Coreference could also be very helpful to gather more training data for the classifier. In the current approach, a partial/full keyword match strategy was employed to detect the entities in the text, discarding many samples in which entities are substituted by pronouns. Although the coreference resolution is far from being solved there are recent promising advances in this field [4]. We leave as a future line of research the application of those techniques and the possible beneficial effects in the performance of the system.

7   Conclusion

In this paper we present our approach to the relation extraction Task 3 of the OKE Challenge 2018. The core of the system is a deep learning engine that extract the relation between a pair of detected entities. The set of relations obtained from a document is fed to another system that applies expert knowledge in order to filter out relations that are illogical or to create new ones when they can be directly implied from other relations. Finding data for training that system is very complicated and much of our effort went in that direction. Although we employed more than 1 million records for training, they were not completely revised by a human and they are far from being ideal. Training records are still very unbalanced and some of them are clearly biased (e.g. existing records of the relation *trainer* in DBPedia are about professional fighting whereas the figure of trainers exists in nearly all sports).

Another improvement of the system could be the inclusion of a coreference system in the analytical loop. On one hand, the identification of multiple references to the same entity in a document will help with the extraction of more training records. On the other hand, it will avoid some mistakes of the system when assigning a relation between two instances of the same entity in a sentence.

The application of a classifier, which acted on each pair of entities individually, and an output control module, which applied a global vision of all the relations extracted from a document, showed a very positive effect on the final score of the system. The research on better ways of obtaining and applying those expert knowledge over weak supervised machine learning models could have many interesting applications in NLP problems (e.g. textual entailment, relation extraction, common sense knowledge acquisition). As they grow in complexity, finding large sized ground truth data for learning and validation is more challenging.

References

1. Auer, Sören, Bizer, Christian, Kobilarov, Georgi, Lehmann, Jens, Cyganiak, Richard, Ives, Zachary: DBpedia: a nucleus for a web of open data. In: Aberer, Karl, Choi, Key-Sun, Noy, Natasha, Allemang, Dean, Lee, Kyung-Il, Nixon, Lyndon, Golbeck, Jennifer, Mika, Peter, Maynard, Diana, Mizoguchi, Riichiro, Schreiber, Guus, Cudré-Mauroux, Philippe (eds.) ASWC/ISWC -2007. LNCS, vol. 4825, pp. 722–735. Springer, Heidelberg (2007). https://doi.org/10.1007/978-3-540-76298-0_52
2. Bordes, A., Usunier, N., Chopra, S., Weston, J.: Large-scale simple question answering with memory networks. arXiv preprint arXiv:1506.02075 (2015)
3. Cannaviccio, M., Barbosa, D., Merialdo, P.: Accurate fact harvesting from natural language text in wikipedia with lector. In: Proceedings of the 19th International Workshop on Web and Databases, p. 9. ACM (2016)

4. Clark, K., Manning, C.D.: Improving coreference resolution by learning entity-level distributed representations. arXiv preprint arXiv:1606.01323 (2016)
5. Durrett, G., Klein, D.: Easy victories and uphill battles in Coreference resolution. In: Proceedings of the 2013 Conference on Empirical Methods in Natural Language Processing, pp. 1971–1982 (2013)
6. Exner, P., Nugues, P.: Entity extraction: from unstructured text to DBpedia RDF triples. In: The Web of Linked Entities Workshop (WoLE 2012), pp. 58–69. CEUR (2012)
7. Hochreiter, S., Schmidhuber, J.: Long short-term memory. Neural Comput. **9**(8), 1735–1780 (1997)
8. Lample, G., Ballesteros, M., Subramanian, S., Kawakami, K., Dyer, C.: Neural architectures for named entity recognition. arXiv preprint arXiv:1603.01360 (2016)
9. Li, X., Taheri, A., Tu, L., Gimpel, K.: Commonsense knowledge base completion. In: Proceedings of the 54th Annual Meeting of the Association for Computational Linguistics (Volume 1: Long Papers), vol. 1, pp. 1445–1455 (2016)
10. Lin, Z., et al.: A structured self-attentive sentence embedding. In: Proceedings of the 5th International Conference on Learning Representations (ICLR) (2017)
11. Mintz, M., Bills, S., Snow, R., Jurafsky, D.: Distant supervision for relation extraction without labeled data. In: Proceedings of the Joint Conference of the 47th Annual Meeting of the ACL and the 4th International Joint Conference on Natural Language Processing of the AFNLP: Volume 2-Volume 2, pp. 1003–1011. Association for Computational Linguistics (2009)
12. Mitchell, T., et al.: Never-ending learning. In: Proceedings of the Twenty-Ninth AAAI Conference on Artificial Intelligence (AAAI-15) (2015)
13. Ostermann, S., Roth, M., Modi, A., Thater, S., Pinkal, M.: Semeval-2018 task 11: Machine comprehension using commonsense knowledge. In: Proceedings of The 12th International Workshop on Semantic Evaluation, pp. 747–757 (2018)
14. Paulus, R., Xiong, C., Socher, R.: A deep reinforced model for abstractive summarization. arXiv preprint arXiv:1705.04304 (2017)
15. Riedel, S., Yao, L., McCallum, A.: Modeling relations and their mentions without labeled text. In: Balcázar, J.L., Bonchi, F., Gionis, A., Sebag, M. (eds.) ECML PKDD 2010. LNCS (LNAI), vol. 6323, pp. 148–163. Springer, Heidelberg (2010). https://doi.org/10.1007/978-3-642-15939-8_10
16. Silva, V.S., Freitas, A., Handschuh, S.: Building a Knowledge Graph from Natural Language Definitions for Interpretable Text Entailment Recognition. ArXiv e-prints (June 2018)
17. Socher, R., Chen, D., Manning, C.D., Ng, A.: Reasoning with neural tensor networks for knowledge base completion. In: Advances in Neural Information Processing Systems, pp. 926–934 (2013)
18. Speer, R., Havasi, C.: Representing general relational knowledge in conceptnet 5. In: LREC, pp. 3679–3686 (2012)
19. Tai, K.S., Socher, R., Manning, C.D.: Improved semantic representations from tree-structured long short-term memory networks. arXiv preprint arXiv:1503.00075 (2015)
20. Usbeck, R., et al.: Gerbil: general entity annotator benchmarking framework. In: Proceedings of the 24th International Conference on World Wide Web, pp. 1133–1143. International World Wide Web Conferences Steering Committee (2015)
21. Wang, L.: Yuanfudao at semeval-2018 task 11: three-way attention and relational knowledge for commonsense machine comprehension. arXiv preprint arXiv:1803.00191 (2018)

22. Weston, J., Bordes, A., Yakhnenko, O., Usunier, N.: Connecting language and knowledge bases with embedding models for relation extraction, pp. 1366–1371 (2013)
23. Wu, Y., et al.: Google's neural machine translation system: bridging the gap between human and machine translation. arXiv preprint arXiv:1609.08144 (2016)

The Scalable Question Answering Over Linked Data Challenge

The Scalable Question Answering Over Linked Data (SQA) Challenge 2018

Giulio Napolitano[1(✉)], Ricardo Usbeck[2], and Axel-Cyrille Ngonga Ngomo[2]

[1] Fraunhofer-Institute IAIS, Sankt Augustin, Germany
`giulio.napolitano@iais.fraunhofer.de`
[2] Data Science Group, University of Paderborn, Paderborn, Germany
`{ricardo.usbeck,axel.ngonga}@uni-paderborn.de`

1 Introduction

Question answering (QA) systems, which source answers to natural language questions from Semantic Web data, have recently shifted from the research realm to become commercially viable products. Increasing investments have refined an interaction paradigm that allows end users to profit from the expressive power of Semantic Web standards, while at the same time hiding their complexity behind intuitive and easy-to-use interfaces. Not surprisingly, after the first excitement we did not witness a cooling-down phenomenon: regular interactions with question answering systems have become more and more natural. As consumers expectations around the capabilities of systems able to answer questions formulated in natural language keep growing, so is the availability of such systems in various settings, devices and languages. Increasing usage in real (non-experimental) settings have boosted the demand for resilient systems, which can cope with high volume demand.

The Scalable Question Answering (SQA) challenge stems from the longstanding Question Answering over Linked Data (QALD)[1] challenge series, aiming at providing an up-to-date benchmark for assessing and comparing state-of-the-art-systems that mediate between a large volume of users, expressing their information needs in natural language, and RDF data. It thus targets all researchers and practitioners working on querying Linked Data, natural language processing for question answering, information retrieval and related topics. The main goal is to gain insights into the strengths and shortcomings of different approaches and into possible solutions for coping with the increasing volume of requests that QA systems have to process, as well as with the large, heterogeneous and distributed nature of Semantic Web data.

The SQA challenge at ESWC2018 was supported by the EU project HOBBIT[2]. The employment of the platform provided by the HOBBIT project guaranteed a robust, controlled setting offering rigorous evaluation protocols. This now popular platform for the benchmarking of linked data has attracted a high

---

[1] http://www.sc.cit-ec.uni-bielefeld.de/qald/.
[2] https://project-hobbit.eu/.

© Springer Nature Switzerland AG 2018
D. Buscaldi et al. (Eds.): SemWebEval 2018, CCIS 927, pp. 69–75, 2018.
https://doi.org/10.1007/978-3-030-00072-1_6

number of users, as semantic data and knowledge bases are gaining relevance also for information retrieval and search, both in academia and for industrial players.

2   Task and Dataset Creation

The key difficulty for Scalable Question Answering over Linked Data is in the need to translate a users information request into such a form that it can be efficiently evaluated using standard Semantic Web query processing and inferencing techniques. Therefore, the main task of the SQA challenge was the following:

> Given an RDF dataset and and a large volume of natural language questions, return the correct answers (or SPARQL queries that retrieves those answers).

Successful approaches to Question Answering are able to scale up to big data volumes, handle a vast amount of questions and accelerate the question answering process (e.g. by parallelization), so that the highest possible number of questions can be answered as accurately as possible in the shortest time. The focus of this task is to withstand the confrontation of the large data volume while returning correct answers for as many questions as possible.

Dataset

Training set. The training set consists of the questions compiled for the award-nominated LC-QuAD dataset [4], which comprises 5000 questions of variable complexity and their corresponding SPARQL queries over DBpedia (2016-10 dump) as the RDF knowledge base[3]. The questions were compiled on the basis of query templates which are instantiated by seed entities from DBpedia into normalised natural question structures. These structures have then been transformed into natural language questions by native speakers. In contrast to the analogous challenge task run at ESWC in 2017 [5], the adoption of this new dataset ensured an increase in the complexity of the questions as the corresponding SPARQL queries are not limited to one triple in the WHERE clause but also include patterns of two or three triples. In the dataset some spelling mistakes and anomalies are also introduced, as a way to simulate a noisy real-world scenario in which questions may be served to the system imperfectly as a result, for instance, of speech recognition failures or typing errors.

Test set. The test set was created by manually paraphrasing 2200 questions from the LC-QuAD dataset. The paraphrased questions where ordered by their Vector Extrema Cosine Similarity score [3] to the original questions and only the first 1830 questions were retained. As in the original source questions, we intentionally left in typos and grammar mistakes in order to reproduce realistic scenarios of imperfect input.

---

[3]http://downloads.dbpedia.org/2016-10/core-i18n/.

Data Format

The test data for the challenge, without the SPARQL queries, can be found in our project repository https://hobbitdata.informatik.uni-leipzig.de/SQAOC/. We used a format similar to the QALD-JSON format[4] and the following sample shows the first two entries of the training set:

```
1  {
2      "dataset": {
3          "id": "lcquad-v1"
4      },
5      "questions": [
6          {
7              "hybrid": "false",
8              "question": [
9                  {
10                     "string": "Which comic characters
                        are painted by Bill Finger?",
11                     "language": "en"
12                 }
13             ],
14             "onlydbo": true,
15             "query": {
16                 "sparql":
17                 "SELECT DISTINCT ?uri
18                 WHERE {?uri <http://dbpedia.org/
                    ontology/creator> <http://
                    dbpedia.org/resource/
                    Bill_Finger> .
19                 ?uri <http://www.w3.org/1999/02
                    /22-rdf-syntax-ns#type> <
                    http://dbpedia.org/ontology
                    /ComicsCharacter>
20                 }"
21             },
22             "aggregation": false,
23             "_id": "f0a9f1ca14764095ae089b152e0e7f12",
24             "id": 0
25         },
26         {
27             "hybrid": "false",
28             "question": [
29                 {
30                     "string": "Was winston churchill
                        the prime minister of Selwyn
                        Lloyd?",
```

---

[4]https://github.com/AKSW/gerbil/wiki/Question-Answering.

```
31              "language": "en"
32          }
33      ],
34      "onlydbo": true,
35      "query": {
36          "sparql":
37              "ASK WHERE {<http://dbpedia.org/
                    resource/Selwyn_Lloyd> <http://
                    dbpedia.org/ontology/
                    primeMinister> <http://dbpedia.
                    org/resource/Winston_Churchill>
                    }"
38      },
39      "aggregation": true,
40      "_id": "30b709079ea5421cb33c227c3feb9019",
41      "id": 1
42  },
43  ...
```

3 Evaluation

The SQA challenge provides an automatic evaluation tool (based on GERBIL QA [6] and integrated into the HOBBIT platform)[5,6] that is open source and available for everyone to re-use. The HOBBIT platform also incorporates a leaderboard feature to facilitate comparable evaluation and result display of systems participating in challenges. The tool is also accessible online, so that participants were able to upload their systems as Docker images and check their (and others') performance via a webservice. The ranking of the systems was based on the usual KPIs (precision, recall and F measure) plus a response power measure, which is also taking into account the ability of the systems to cope with high volume demand without failure. The response power is the harmonic mean of three measures: precision, recall and the ratio between processed questions (an empty answer is considered as processed, a missing answer is considered as unprocessed) and total number of questions sent to the system. The final ranking was on

1. response power
2. precision
3. recall
4. F measure

[5] http://gerbil-qa.aksw.org/gerbil/.
[6] http://master.project-hobbit.eu/.

in that order. For each system q, precision, recall and response power are computed as follows:

$$precision(q) = \frac{\text{number of correct system answers for } q}{\text{number of system answers for } q}$$

$$recall(q) = \frac{\text{number of correct system answers for } q}{\text{number of gold standard answers for } q}$$

$$\text{response power}(q) = \frac{3}{\frac{1}{precision(q)} + \frac{1}{recall(q)} + \frac{processed}{submitted}}$$

The benchmark sends to the QA system one question at the start, two more questions after one minute and continues to send $n+1$ new questions after n minutes. One minute after the last set of questions is dispatched, the benchmarks closes and the evaluation is generated as explained above. The 1830 questions in the dataset allow the running of the benchmark for one hour but for the SQA challenge we limited to 30 sets of questions.

4   Participating Systems

Three teams participated in the SQA challenge. We provide here brief descriptions, please refer to the respective full papers (where they exist) for more detailed explanations.

WDAqua-core1 [1] is built on a rule-based system using a combinatorial approach to generate SPARQL queries from natural language questions. In particular, the system abstracts from the specific syntax of the question and relies on the semantics encoded in the underlying knowledge base. It can answer questions on a number of Knowledge Bases, in different languages, and does not require training.

LAMA [2] was originally developed for QA in French. It was extended for the English language and modified to decompose complex queries, with the aim of improving performance on such queries and reduce response times. The question type (e.g. *Boolean* or *Entity*) is classified by pattern matching and processes by the relevant component to extract entities and properties. Complex questions are decomposed in simple queries by keyword matching.

GQA [7], the Grammatical Question Answering system, is built around a functional programming language with categorial grammar formalism. The question is parsed according to the grammar and the best parse is selected. Finally, this is decomposed into its elements, starting from the innermost, while requests are sent to DBpedia to find the corresponding values and the final answers.

5   Results

The experimental data for the SQA challenge over the test dataset can be found at the following URLs:

- WDAqua: https://master.project-hobbit.eu/experiments/1527792517766,
- LAMA: https://master.project-hobbit.eu/experiments/1528210879939,
- GQA): https://master.project-hobbit.eu/experiments/1528283915360.

By providing human- and machine-readable experimental URIs, we provide deeper insights and repeatable experiment setups.

Note also that the numbers reported here may differ from the publications of the participants, as these figures were not available at the time of participant paper submission (Table 1).

Table 1. Overview of SQA results.

Test	WDAqua	LAMA	GQA
Response power	0.472	0.019	0.028
Micro precision	0.237	0.054	0.216
Micro recall	0.055	0.001	0.002
Micro F1-measure	0.089	0.001	0.004
Macro precision	0.367	0.010	0.018
Macro recall	0.380	0.016	0.019
Macro F1-measure	0.361	0.011	0.019

6   Summary

The Scalable Question Answering over Linked Data challenge introduced a new metric (Response Power) to evaluate the capability of a QA system to perform under increasing stress. For the first time, it also partially employed complex and non-well-formed natural language questions, to make the challenge even closer to real use scenarios. In this challenge, we also kept last year underlying evaluation platform (HOBBIT) based on docker, to account for the need for comparable experiments via webservices. This introduces an entrance threshold for participating teams but ensures a long term comparability of the system performance and a fair and open challenge. Finally, we offered leader boards prior to the actual challenge in order to allow participants to see their performance in comparison with the others. Overall, we are confident that the HOBBIT platform will be able to provide QA challenge support for a long time, making comparable and repeatable question answering research possible.

Acknowledgments. This work was supported by the European Union's H2020 research and innovation action HOBBIT under the Grant Agreement number 688227.

References

1. Diefenbach, D., Singh, K., Maret, P.: On the scalability of the QA system WDAqua-core1. In: Recupero, D.R., Buscaldi, D. (eds.) Semantic Web Challenges, Cham. Springer International Publishing (2018)
2. Radoev, N., Tremblay, M., Zouaq, A., Gagnon, M.: LAMA: a language adaptive method for question answering. In: Recupero, D.R., Buscaldi, D. (eds.) Semantic Web Challenges, Cham. Springer International Publishing (2018)
3. Sharma, S., El Asri, L., Schulz, H., Zumer, J.: Relevance of unsupervised metrics in task-oriented dialogue for evaluating natural language generation. In: CoRR, arXiv:abs/1706.09799 (2017)
4. Trivedi, P., Maheshwari, G., Dubey, M., Lehmann, J.: LC-QuAD: A Corpus for Complex Question Answering over Knowledge Graphs, pp. 210–218. Springer International Publishing, Cham (2017)
5. Usbeck, R., Ngomo, A.-C.N., Haarmann, B., Krithara, A., Röder, M., Napolitano, G.: 7th open challenge on question answering over linked data (QALD-7). In: Dragoni, M., Solanki, M., Blomqvist, E. (eds.) SemWebEval 2017. CCIS, vol. 769, pp. 59–69. Springer, Cham (2017). https://doi.org/10.1007/978-3-319-69146-6_6
6. Usbeck, R., et al.: Benchmarking question answering systems. Semant. Web J. (2018)
7. Zimina, E., et al.: GQA: grammatical question answering for RDF data. In: Recupero, D.R., Buscaldi, D. (eds.), Semantic Web Challenges, Cham. Springer International Publishing (2018)

On the scalability of the QA System WDAqua-core1

Dennis Diefenbach[(✉)], Kamal Singh, and Pierre Maret

Laboratoire Hubert Curien, Saint Etienne, France
{dennis.diefenbach,kamal.singh,pierre.maret}@univ-st-etienne.fr

Abstract. Scalability is an important problem for Question Answering (QA) systems over Knowledge Bases (KBs). Current KBs easily contain hundreds of millions of triples and all these triples can potentially contain the information requested by the user.

In this publication, we describe how the QA system WDAqua-core1 deals with the scalability issue. Moreover, we compare the scalability of WDAqua-core1 with existing approaches.

Keywords: Question answering · Knowledge bases · Scalability WDAqua-core1

1 Introduction

Question Answering (QA) over Knowledge Bases (KBs) is a field in computer science that tries to build a system able to search in a KB the information requested by a user using natural language. For example if a user asks: "What is the capital of Congo?" a QA system over Wikidata should be able to generate the following SPARQL query that retrieves the desired answer:

```
PREFIX wde: <http://www.wikidata.org/entity/>
PREFIX wdp: <http://www.wikidata.org/prop/direct/>
SELECT ?o where {
    wde:Q974 wdp:P36 ?o .
}
```

Scalability refers to the problem of answering a question in a time that is acceptable by the user. This problem mainly depends on the size of the data to query. Note that a KB like Wikidata contains more than 2 billion triples, which corresponds to roughly 250 Gb in an ntriples dump. On the other side, a user nowadays expects response times of around 1 s. Since a QA system is running on a server, in addition to the time required to compute the answer, it is also necessary to add the overhead of the network requests, and the retrieval and rendering of the information related to the answer. This means that very easily scalability becomes a problem for QA systems.

In this publication, we describe how the QA system WDAqua-core1 tackles this problem.

© Springer Nature Switzerland AG 2018
D. Buscaldi et al. (Eds.): SemWebEval 2018, CCIS 927, pp. 76–81, 2018.
https://doi.org/10.1007/978-3-030-00072-1_7

Fig. 1. Screenshot of Trill [3] using WDAqua-core1 as a back-end QA system for the question "What is the capital of Congo?". The answer is given in 0.974 s. In this case Wikidata is queried which is roughly 250 Gb ntriples dump.

2   Approach

In this section, we are going to describe what is the main idea behind the algorithm used by WDAqua-core1. We would describe it using an example. Imagine the QA system receives the question: "What is the capital of Congo?". WDAqua-core1 takes into consideration every n-gram in the question and maps it to potential meanings. For example "what is" can correspond to "What Is... (album by Richie Kotzen)" and capital can refer to the expected property "capital" but also to "Capital (German business magazine)", "Capital Bra (German rapper)" and so on. For this example question, 76 different meanings are taken into consideration. All these meanings are used to generate SPARQL queries that are possible interpretations of the questions. For this example question, we generate 292 possible interpretations. Once all interpretations are generated, they are ranked and the first ranked query is executed and the answer is retrieved. The approach clearly needs to be implemented efficiently to achieve acceptable time performance. For further details, we refer to [4]. A running demo of the approach can be found at:

www.wdaqua.eu/qa

A screen-shot of Trill [3], a reusable front-end for QA systems, using in the back-end WDAqua-core1 can be found in Fig. 1.

3 Scalability

In this section, we are going to describe which are the key elements that improve the scalability of the approach.

- To find the potential meanings of all n-grams in a question we rely on a Lucene Index of the labels of the targeted KB. It is characterized by high efficiency and low memory footprint.
- To construct all possible SPARQL queries out of the potential meanings we use an efficient algorithm. It is described in [4]. To achieve this we rely on HDT (Header Dictionary Triples) as an index. HDT is used because it allows fast breath search operations over RDF graphs. In fact RDF graphs are stored in HDT like adjecency lists. This is an ideal data structure to perform breath search operations. HDT is also characterized by low memory footprint.
- The current implementation uses as few HTTP requests as possible, to reduce the overhead they generate.

Above allows us to run a QA system over DBpedia and Wikidata with a memory footprint of only 26 Gb. Above ideas in particular allow the system to scale horizontally with the size of a given KB.

4 Experiments

Table 1 summarizes the existing QA solutions evaluated over the QALD benchmarks. It also indicates the average run time of all the queries per given task, if indicated in the corresponding publication. This clearly depends on the underlying hardware, but independently from that, it provides a hint on the scalability of the proposed approach. Note that, for most of the systems, no information is provided about the performance in terms of execution time. One of the main reasons is that these QA systems do not exist as one pipeline, but often authors write sub-results of sub-pipelines into files and arrive step-wise at the final answer. For the few systems that provided the performance in terms of execution time, this varies, depending on the question, from around 1 s for gAnswer [20] to more than 100 s for some questions in the case of SINA [15]. WDAqua-core1 obtains results similar to the best systems. It is an entire pipeline and available on line.

Table 1. Table comparing WDAqua-core1 with other QA systems evaluated over QALD-3 (over DBpedia 3.8), QALD-4 (over DBpedia 3.9), QALD-5 (over DBpedia 2014), QALD-6 (over DBpedia 2015-10), QALD-7 (over DBpedia 2016-04). We indicated the average running times of a query if the corresponding publication contained the information. These evaluation are performed on different hardware, but still give a good idea about scalability.

QA system	P	R	F	Time	QA system	P	R	F	Time
QALD-3					QALD-5				
WDAqua-core1	**0.58**	**0.46**	**0.51**	**1.08s**	Xser [17]	0.74	0.72	0.73	-
gAnswer [20]∗	0.40	0.40	0.40	0.971 s	**WDAqua-core1**	**0.56**	**0.41**	**0.47**	**0.62 s**
RTV [8]	0.32	0.34	0.33	-	AskNow[7]	0.32	0.34	0.33	-
Intui2 [5]	0.32	0.32	0.32	-	QAnswer[14]	0.34	0.26	0.29	-
SINA [15]∗	0.32	0.32	0.32	≈10–20 s	SemGraphQA[2]	0.19	0.20	0.20	-
DEANNA [18]∗	0.21	0.21	0.21	≈1–50 s	YodaQA[1]	0.18	0.17	0.18	-
SWIP [13]	0.16	0.17	0.17	-	QuerioDali[11]		0.48	?	?
Zhu et al. [19]∗	0.38	0.42	0.38	-					
QALD-4					QALD-6				
Xser [17]	0.72	0.71	0.72	-	UTQA [16]	0.82	0.69	0.75	-
WDAqua-core1	**0.56**	**0.30**	**0.39**	**0.46 s**	UTQA [16]	0.76	0.62	0.68	-
gAnswer [20]	0.37	0.37	0.37	0.972 s	UTQA [16]	0.70	0.61	0.65	-
CASIA [10]	0.32	0.40	0.36	-	**WDAqua-core1**	**0.62**	**0.40**	**0.49**	**0.93 s**
Intui3 [6]	0.23	0.25	0.24	-	SemGraphQA [2]	0.70	0.25	0.37	-
ISOFT [12]	0.21	0.26	0.23	-	QALD-7				
Hakimov [9]∗	0.52	0.13	0.21	-	**WDAqua-core1**	**0.63**	**0.32**	**0.42**	**0.47 s**

5 Conclusion

We have described the scalability problem of QA system over KB. Moreover, we have described how scalability is addressed in WDAqua-core1. Finally, we have compared the runtime performance of WDAqua-core1 with existing QA solutions evaluated over the popular QALD benchamrk series. The presented results show that WDAqua-core1 has a runtime performance that can compete with all evaluated system.

Note: There is a Patent Pending for the presented approach. It was submitted the 18 January 2018 at the EPO and has the number EP18305035.0.

Acknowledgments. Parts of this work received funding from the European Union's Horizon 2020 research and innovation programme under the Marie Skodowska-Curie grant agreement No. 642795, project: Answering Questions using Web Data (WDAqua).

References

1. Baudiš, P., Šedivỳ, J.: QALD challenge and the YodaQA system: prototype notes
2. Beaumont, R., Grau, B., Ligozat, A.L.: SemGraphQA@QALD-5: LIMSI participation at QALD-5@CLEF. CLEF (2015)
3. Diefenbach, D., Amjad, S., Both, A., Singh, K., Maret, P.: Trill: a reusable frontend for QA systems. In: ESWC P&D (2017)
4. Diefenbach, D., Both, A., Singh, K., Maret, P.: Towards a question answering system over the semantic web (2018), arXiv:1803.00832
5. Dima, C.: Intui2: a prototype system for question answering over linked data. In: Proceedings of the Question Answering over Linked Data lab (QALD-3) at CLEF (2013)
6. Dima, C.: Answering natural language questions with Intui3. In: Conference and Labs of the Evaluation Forum (CLEF) (2014)
7. Dubey, M., Dasgupta, S., Sharma, A., Höffner, K., Lehmann, J.: AskNow: a framework for natural language query formalization in SPARQL. In: Sack, H., Blomqvist, E., d'Aquin, M., Ghidini, C., Ponzetto, S.P., Lange, C., (eds.) ESWC 2016. LNCS, vol. 9678, pp. 300–316. Springer, Cham (2016). https://doi.org/10.1007/978-3-319-34129-3_19
8. Giannone, C., Bellomaria, V., Basili, R.: A HMM-based approach to question answering against linked data. In: Proceedings of the Question Answering over Linked Data lab (QALD-3) at CLEF (2013)
9. Hakimov, S., Unger, C., Walter, S., Cimiano, P.: Applying semantic parsing to question answering over linked data: addressing the lexical gap. In: Biemann, C., Handschuh, S., Freitas, A., Meziane, F., Métais, E. (eds.) NLDB 2015. LNCS, vol. 9103, pp. 103–109. Springer, Cham (2015). https://doi.org/10.1007/978-3-319-19581-0_8
10. He, S., Zhang, Y., Liu, K., Zhao, J.: CASIA@ V2: a MLN-based question answering system over linked data. In: Proceedings of QALD-4 (2014)
11. Lopez, V., Tommasi, P., Kotoulas, S., Wu, J.: QuerioDALI: question answering over dynamic and linked knowledge graphs. In: Groth, P., Simperl, E., Gray, A., Sabou, M., Krötzsch, M., Lecue, F., Flöck, F., Gil, Y. (eds.) ISWC 2016. LNCS, vol. 9982, pp. 363–382. Springer, Cham (2016). https://doi.org/10.1007/978-3-319-46547-0_32
12. Park, S., Shim, H., Lee, G.G.: ISOFT at QALD-4: Semantic similarity-based question answering system over linked data. In: CLEF (2014)
13. Pradel, C., Haemmerlé, O., Hernandez, N.: A semantic web interface using patterns: The SWIP system. In: Croitoru, M., Rudolph, S., Wilson, N., Howse, J., Corby, O. (eds.) GKR 2011. LNCS (LNAI), vol. 7205, pp. 172–187. Springer, Heidelberg (2012). https://doi.org/10.1007/978-3-642-29449-5_7
14. Ruseti, S., Mirea, A., Rebedea, T., Trausan-Matu, S.: QAnswer-Enhanced Entity Matching for Question Answering over Linked Data. CLEF (2015)
15. Shekarpour, S., Marx, E., Ngomo, A.C.N., Auer, S.: Sina: semantic interpretation of user queries for question answering on interlinked data. Web Semantics: Science, Services and Agents on the World Wide Web 30 (2015)
16. Pouran-ebn veyseh, A.: Cross-lingual question answering using profile HMM & unified semantic space. In: ESWC (2016, to appear)
17. Xu, K., Feng, Y., Zhao, D.: Xser@ QALD-4: Answering Natural Language Questions via Phrasal Semantic Parsing (2014)

18. Yahya, M., Berberich, K., Elbassuoni, S., Ramanath, M., Tresp, V., Weikum, G.: Natural language questions for the web of data. In: Proceedings of the 2012 Joint Conference on Empirical Methods in Natural Language Processing and Computational Natural Language Learning. Association for Computational Linguistics (2012)
19. Zhu, C., Ren, K., Liu, X., Wang, H., Tian, Y., Yu, Y.: A graph traversal based approach to answer Non-aggregation questions over DBpedia (2015), arXiv preprint arXiv:1510.04780
20. Zou, L., Huang, R., Wang, H., Yu, J.X., He, W., Zhao, D.: Natural language question answering over RDF: a graph data driven approach. In: Proceedings of the 2014 ACM SIGMOD International Conference on Management of data. ACM (2014)

GQA: Grammatical Question Answering
for RDF Data

Elizaveta Zimina, Jyrki Nummenmaa$^{(\boxtimes)}$, Kalervo Järvelin, Jaakko Peltonen,
Kostas Stefanidis, and Heikki Hyyrö

University of Tampere, Tampere, Finland
{elizaveta.zimina,jyrki.nummenmaa,kalervo.jarvelin,jaakko.peltonen,
kostas.stefanidis,heikki.hyyro}@uta.fi

Abstract. Nowadays, we observe a rapid increase in the volume of RDF
knowledge bases (KBs) and a need for functionalities that will help users
access them in natural language without knowing the features of the KBs
and structured query languages, such as SPARQL. This paper introduces
Grammatical Question Answering (GQA), a system for answering questions in the English language over DBpedia, which involves parsing of
questions by means of Grammatical Framework and further analysis of
grammar components. We built an abstract conceptual grammar and a
concrete English grammar, so that the system can handle complex syntactic constructions that are in the focus of the SQA2018 challenge. The
parses are further analysed and transformed into SPARQL queries that
can be used to retrieve the answers for the users' questions.

Keywords: Grammatical framework · DBpedia
Question answering · SQA

1 Introduction

Retrieving information from knowledge bases is a commonplace task, which
is becoming further widespread due to digitalisation of society; this has also
increased the need for systems that support natural language interaction.

Typically, data in knowledge bases are represented using the RDF model.
In RDF, everything we wish to describe is a resource that may be a person,
an institution, a thing, a concept, or a relation between other resources. The
building block of RDF is a triple of the form (*subject, predicate, object*). The
flexibility of the RDF data model allows representation of both schema and
instance information in the form of RDF triples.

The traditional way to retrieve RDF data is through SPARQL [26], the W3C
recommendation language for querying RDF datasets. SPARQL queries are built
from triple patterns and determine the pattern to seek for; the answer is the
part(s) of the set of RDF triples that match(es) this pattern. The correctness
and completeness of answers of SPARQL queries are key challenges. SPARQL is
a structured query language that allows users to submit queries that precisely

© Springer Nature Switzerland AG 2018
D. Buscaldi et al. (Eds.): SemWebEval 2018, CCIS 927, pp. 82–97, 2018.
https://doi.org/10.1007/978-3-030-00072-1_8

identify their information needs, but requires them to be familiar with the syntax, and the complex semantics of the language, as well as with the underlying schema or ontology. Moreover, this interaction mode assumes that users are familiar with the content of the knowledge base and also have a clear understanding of their information needs. As databases become larger and accessible to a more diverse and less technically oriented audience, new forms of data exploration and interaction become increasingly attractive to aid users navigate through the information space and overcome the challenges of information overload [24,25].

In this work, we present the Grammatical Question Answering (GQA) system for answering questions in the English language over DBpedia 2016-04 [8]. The GQA system's key technology is Grammatical Framework (GF) [23], which provides parsing of questions and extraction of conceptual categories, the main types of which are Entity, Property, Verb Phrase, Class, and Relative Clause. The system "unfolds" the parse layer by layer and matches its components with the KB schema and contents to formulate SPARQL queries corresponding to the English questions. The GQA was tested on the SQA2018 training set [15] and showed high precision in answering both simple and complex questions.

The rest of the paper is structured as follows. Section 2 presents an overview of the GQA system and resources used, while Sect. 3 describes the GQA grammar. Section 4 provides the details of the parse interpretation module, and Sect. 5 analyses the system's testing results. The related work is surveyed in Sect. 6, and conclusions and perspectives for improvement are stated in Sect. 7.

2   GQA Modules and Resources

2.1   General Architecture of GQA

The keys to successful question answering over knowledge bases are correct interpretation of questions and proper retrieval of information. In the GQA system (Fig. 1), this is realised by means of the two main modules:

• the Conceptual Parsing module, which obtains all possible parses of a question according to the GQA grammar and selects the most suitable one, and

• the Parse Interpretation module, which analyses the question parse, gradually "unfolding" its elements and making requests to the knowledge base in order to control the process of parse interpretation and finally arrive at the most probable answer(s).

The parsing module exploits the technology of Grammatical Framework, and the GQA grammar is built on top of the existing GF Resource Grammar Library. The parse analysis is performed in Python, and the KB is indexed by means of Apache Lucene.

2.2   Grammatical Framework

Grammatical Framework (GF) [22,23] is a functional programming language and a platform for multilingual grammar-based applications. A GF grammar

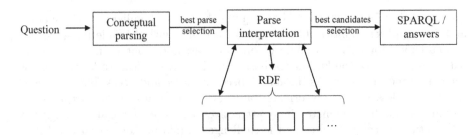

Fig. 1. General architecture of GQA

consists of an abstract syntax, which declares the grammar's meanings and their correlation in the form of abstract syntax trees, and one or more concrete syntaxes, which determine the mapping of abstract syntax trees onto strings and vice versa.

The analysis of natural language strings is realised through parsing, that is, deriving an abstract syntax tree (or several) from a string. Conversely, strings can be generated by means of linearisation of parse trees.

Natural language grammars implemented in GF are collected in the GF Resource Grammar Library (RGL) [5]. The concrete grammars for all (currently 40) languages in the RGL comply with the common abstract syntax, which aims to reflect the universal linguistic categories and relations.

Thus, in most cases the RGL categories correspond to parts of speech (noun, verb, etc.) and textual units (sentence, question, text, etc.). These categories can act as arguments to functions forming other categories. For example, we can define (in a simplified way) an abstract syntax rule that takes an pronoun (PN) and a verb (V) and forms a sentence (S) as follows:

```
mkS : PN -> V -> S.
```

Let us write the linearisation definition of this rule for an English concrete syntax, taking only the present simple tense. To make the correct agreement possible, we first need to specify the *Person* and *Number* parameters:

```
Person = Fisrt | Second | Third ;
Number = Sg | Pl ;
```

and then can define the linearisation types for *NP*, *V* and *S*:

```
PN = {s : Str ; p : Person ; n : Number} ;
V  = {s : Person => Number => Str} ;
S  = {s : Str}.
```

Any *PN* has a permanent string value s as well as inherent values p and n, standing for *Person* and *Number*. The string value s of V is represented by an inflectional table, so that it can change depending on the *Person* and *Number* values. S has only a simple string value s.

Finally, the linearisation definition for the rule *mkS* can be written as:

```
mkS pron verb = {s = pron.s ++ verb.s ! pron.p ! pron.n}.
```

The string value s of S is obtained by joining the generated string values of *PN* and *V* so that the verb agrees with the pronoun in *Person* and *Number*.

We can now add some vocabulary rules to the abstract syntax:

```
we_PN, you_PN, she_PN : PN ;
sleep_V : V ;
```

and their linearisation definitions to the concrete syntax:

```
we_PN    = {s = "we"  ; p = First  ; n = Pl} ;
you_PN   = {s = "you" ; p = Second ; n = Pl} ;
she_PN   = {s = "she" ; p = Third  ; n = Sg} ;
sleep_V  = {s = table {
             First  => table {Sg => "sleep"  ; Pl => "sleep"} ;
             Second => table {Sg => "sleep"  ; Pl => "sleep"} ;
             Third  => table {Sg => "sleeps" ; Pl => "sleep"}}} ;
```

So the parse trees of the strings *we sleep, you sleep* and *she sleeps* will look like:

```
mkS we_PN sleep_V ;
mkS you_PN sleep_V ;
mkS she_PN sleep_V.
```

2.3 DBpedia

The GQA system exploits the English DBpedia version 2016-04 [8] that involves: entity labels (6.0 M), redirect pages having alternative labels of entities and leading to pages with "canonical" labels (7.3 M), disambiguation pages containing entities with multiple meanings (260 K), infobox statements (triples where properties are extracted from infoboxes of Wikipedia articles, often noisy and unsystematic; 30.0 M), mapping-based statements (triples taken from the hand-generated DBpedia ontology, cleaner and better structured than the infobox dataset; 37.5 M), and instance type statements attributing an entity to one or several DBpedia ontology types, e.g. *person, country, book*, etc. (36.7 M).

To be able to access this large volume of information quickly, we built an index of all necessary data components by means of Apache Lucene [2]. For infobox and mapping-based statements, we also make inverse indexes, so that the search can be done either by subjects or by values of RDF triples.

3 The GQA Grammar

3.1 Overview

The GQA grammar complies with the morphological principles underlying the RGL and at the same time operates with new conceptual categories that make it possible to classify questions and to reveal their components and relations between them. The main conceptual categories in the GQA grammar include:

- *Entity*: usually a proper name, e.g. *In which country does the **Nile**[1] start?*
- *Property*: a noun phrase that coincides with the label of some property in DBpedia, e.g. *What is the **area code** of Berlin?*
- *VPChunk*: corresponds to all possible morphological forms of *VP* (verb phrase) and usually represents a reformulation of some property (e.g. the verb *dissolve* in the *when*-question *When did the Ming dynasty **dissolve**?* indicates the *dissolution date* property),
- *Class*: a noun phrase that corresponds to an ontology class in DBpedia, e.g. *List all the **musicals** with music by Elton John.*
- *RelCl*: a relative clause, e.g. *Who owns the company **that made the Edsel Villager**?*
- *Q*: the starting category comprising all kinds of questions learnt through the analysis of the training sets. For example, the question *When did Operation Overlord commence?* will be recognised by the function *WhenDidXVP* with the definition *Entity* → *VPChunk* → *Q*.

3.2   Simple Entity

So-called simple entities in the GQA grammar are the only elements that are not actually coded, but are rather regarded as symbols, that is, lists of words separated by whitespaces in natural language questions. Thus, simple entities are formed by means of functions:

oneWordEnt : String → *Entity ;*
twoWordEnt : String → *String* → *Entity ;*
up to *tenWordEnt*.

The English DBpedia 2016-04 comprises 13.3M labels, i.e. names of entities and their redirects (names of pages redirecting to the main pages, e.g. *Houses of Parliament* → *Palace of Westminster*). Indexing the inverse properties (described in Sect. 3.3) reveals even more entities without separate pages in DBpedia. It was impossible and in fact not necessary to code such a number of entities in the grammar, since it would dramatically slow the parsing down. Instead, we built an index of all DBpedia labels and use it for finding entity links. In many cases the input phrase needs some modification before looking for it in the index. These modifications include:

- removal of the article in the beginning of the phrase (*the Isar* → http:// dbpedia.org/resource/Isar),
- capitalisation of all words, excluding auxiliary parts of speech (*maritime museum of San Diego* → http://dbpedia.org/resource/Maritime_Museum_of_San_Diego),
- rearrangement of components separated by the *in* preposition (*lighthouse in Colombo* → http://dbpedia.org/resource/Colombo_Lighthouse).

Quite often the label found in the index leads us to the disambiguation page containing links of several possible entities. For example, the entity *Sherlock* in

---

[1] Words belonging to the category/rule under consideration are given in bold.

the question *What company involved in the development of Sherlock did Arthur Levinson work for?* is highly polysemous (20 readings in DBpedia), and we should check all of them to find the one that matches all properties expressed in the question (http://dbpedia.org/resource/Sherlock_(software)).

3.3   Property

Infobox and Ontology Properties. We collected the labels of all properties in the 2016-04 version of DBpedia. This yielded almost 17,000 grammar rules, most of which are due to DBpedia's noisiness (e.g. properties *cj*, *ci*, or *ch*) and are of no use for question answering. This set still contains "valid" properties (e.g. *height* or *spouse*) and is not large enough to slow down parsing significantly.

Search is performed among property links that start either with http://dbpedia.org/property/ (Wikipedia infobox properties) or http://dbpedia.org/ontology/ (better structured ontology based on the most commonly used infoboxes). In the GQA grammar property name endings correspond to the function names with suffixes _O (e.g. *owner_O* in the above example) or _P showing whether *ontology* or *property* is used in the full link. As a result, properties often duplicate each other, and the correct answer can be obtained with any of them (e.g. http://dbpedia.org/ontology/birthPlace or http://dbpedia.org/property/birthPlace). Thus, the best parse of the question *What is the birthplace of Louis Couperin?* will contain the function *birthPlace_O*, since _O functions are given higher priority in selecting the best parse, but the http://dbpedia.org/property/birthPlace property will be also checked if no answer is found with the ontology property.

At the same time, some infobox properties do not have corresponding ontology properties (e.g. http://dbpedia.org/property/placeOfBirth), but are identical in meaning with other properties and in some cases can be the only key to the answer. Analysing the training datasets, we develop a separate database *Functions* with groups of identical properties, each group stored under the name of the rule that would be most likely selected in the best parse.

Inverse Properties. The question type *What is the [property] of [entity]?* assumes that after the resolution of the entity we need to look for the value of the corresponding property that should be present in this page. A number of syntactic and semantic constructions require so-called inverse search, when we look for a subject, not a value in a triple. For example, the SPARQL query for the question *Whose deputy is Neil Brown?* looks like

```
SELECT DISTINCT ?uri WHERE {?uri <http://dbpedia.org/property/
deputy> <http://dbpedia.org/resource/Neil_Brown_(Australian_
politician)>}.
```

Since the genitive construction is used, the target variable *?uri* is in the position of the triple subject.

We built a separate "inverse" index, which makes this kind of search possible. In our *Functions* database we assign a Boolean value to each property, depending on the function that calls it and showing whether the direct (*False*) or inverse (*True*) search is needed. The value *True* is mostly assigned to properties assumed in participle clauses with past participles constructions and some verb phrases, described in Sect. 3.5.

3.4   Class

A DBpedia ontology class can be considered as a generic term that an entity belongs to (e.g. *written work, beverage, holiday*, etc). As well as ontology property URIs, class URIs have the prefix http://dbpedia.org/ontology/. Classes in questions can have various syntactic roles and forms, e.g. in English they can be inflected for case and number. For example, the parses of the questions *Which magazine's editor is married to Crystal Harris?* and *Give me all magazines whose editors live in Chicago* both should contain the function *Magazine_Class*, so that the parse analysis module will "know" that the search should be done among the entities belonging to the class with the URI http://dbpedia.org/ontology/Magazine.

This morphological flexibility of classes is obtained by means of the RGL paradigms. We collected 764 classes currently existing in the DBpedia ontology and parsed them as common noun chunks, which can take any morphological form. For example, in the GQA grammar the *Magazine_Class* function is linearised as

```
Magazine_Class = UseN magazine_N.
```

The linearisation of *magazine_N* is taken from the RGL English wide-coverage dictionary. This noun type complies with the inflection table

```
Sg => table {Nom => magazine ; Gen => magazine + "'s"} ;
Pl => table {Nom => magazine + "s" ; Gen => magazine + "s'"},
```

building the singular and plural noun forms in the nominative and genitive cases.

The function *UseN* creates a common noun out of a noun ($UseN : N \rightarrow CN$), which has the same type definition as *Class*. Finally, any grammatical form of *Class* is detected as a class chunk :

```
Class_Chunk : Class -> ClassChunk
Class_Chunk cl = {s = variants {cl.s ! Sg ! Nom; cl.s ! Pl ! Nom;
                                cl.s ! Sg ! Gen; cl.s ! Pl ! Gen}},
```

so that *magazine, magazine's, magazines* and *magazines'* in a question parse will be represented by the branch Class_Chunk Magazine_Class. In this way, the RGL makes it simpler to focus on the semantic structure of a question, without a great effort in grammar analysis.

Studying the training sets, we revealed a number of reformulations of class names. For example, the *television show* class can be meant if a question contains words like *program, series, show, sitcom, television program, television series*, etc. At the same time, the word *program* can also refer to another class – *radio program*. As with the *Functions* database for ambiguous properties, we built the *Classes* database containing groups of related classes and their textual expressions. The largest groups are gathered under the words *league* (*sports league, American football league, Australian football league*, etc. – 27 options) and *player* (*soccer player, basketball player, ice hockey player*, etc. – 26 options). Currently, the system checks all the classes of the group and selects the largest set of answers belonging to the same class. Those class linearisations that are highly ambiguous significantly decrease the speed of finding the answer.

3.5 Verb Phrase and Participle

Questions can refer to a property in several ways, one of which is question predicates expressed by verbs and verb phrases. For example, in the question *Who owns Torrey Pines Gliderport?* the verb *owns* should be attributed to the property *owner*. To enable our system to do this, the GQA grammar should contain the verb *own* and the parse interpretation module should "know" how to map certain verbs with corresponding properties.

Analysing the training set, we extracted verbs and verb phrases from questions and correlated them with corresponding properties in SPARQL queries. These correlations are recorded in the *Functions* database, which is used at the parse interpretation step.

In the GQA grammar, verbs and verb phrases often act as verb phrase chunks (*VPChunk*), so that they can be used in any morphological form, in a similar way as class chunks described in Sect. 3.4. Thus, *own, owns, owned*, etc. will be recognised by the grammar under the rule *own_V2*[2], due to the verb paradigms inherited from the RGL grammar.

In many cases one and the same verb can refer to different properties. For example, judging by the SPARQL query for the question *Which company owns Evraz?*, we should attribute the verb *own* also to the property *owningCompany*. All alternative readings are stored in the *Functions* database under the corresponding rule, e.g.

```
"own_V2": [("http://dbpedia.org/ontology/owner", False),
            ("http://dbpedia.org/ontology/owningOrganisation", False),
            ("http://dbpedia.org/ontology/owningCompany", False),
            ("http://dbpedia.org/property/owners", False),
            ("http://dbpedia.org/ontology/parentCompany", False)...]
```

---

[2] Following the logic of the RGL grammar, in order to form a *VPChunk*, a *V2* (two-place verb) is turned into the intermediate category *VPSlashChunk* (*V2_to_VPSlash : V2 → VPSlashChunk*) and then is complemented by a noun phrase, in our case – *Entity* (*VPSlash_to_VPChunk : VPSlashChunk → Entity → VPChunk*). A *VPSlashChunk* can be also formed by a participial construction (e.g. *BeDoneBy_VPslash : Participle2 → VPSlashChunk*).

The *False* value means that the corresponding property does not involve inverse search (see Sect. 3.3). At the parse interpretation step the system checks all properties that the function refers to.

Examples of more complex verb phrases include: *originate in, work for, lead to the demise of, do PhD under*, etc. In the GQA grammar they are referred as *idiomatic VPSlashChunks (IVPS)*.

Similarly to verb phrases, participial constructions (e.g. *located at, followed by, starring in*, etc.), are also learnt from the training sets and stored in the *Functions* database. An exception is the most common model *past participle + by*, when the verb that the participle is derived from already exists in *Functions*, e.g. *written by* in the question *What are the movies written by Nick Castle?* The properties that *written by* refers to are the same as those of the verb *write*, except that we need to carry out the inverse search, since now the author is known and the objects in question are subjects of RDF triples, not values, as with the verb.

3.6   Relative Clause

Below are examples of common rules building relative clauses in the GQA grammar. Words belonging to elements of relative clauses are in square brackets:

- *who/that/which + verb phrase*
  `WhoVP_relcl : VPChunk -> RelCl`
  *Who are the people **who [influenced the writers of Evenor]**?*
- *whose + property + is/are + entity*
  `WhosePropIsX_relcl : Property -> Entity -> RelCl`
  *What is the city **whose [mayor] is [Giorgos Kaminis]**?*
- *whose + property + verb phrase*
  `WhosePropVP_relcl : Property -> VPChunk -> RelCl`
  *List the movies **whose [editors] [are born in London]**.*
- *where + entity + verb phrase*
  `WhereXVP_relcl : Entity -> VPChunk -> RelCl`
  *Where was the battle fought **where [2nd Foreign Infantry Regiment] [participated]**?*

3.7   Complex Entity

Some complex constructions can perform the grammatical functions of entities. The most important of them in the GQA grammar are:

- homogeneous simple entities connected with the conjunction *and*. They are built through the rule:
  `EntityAndEntity : Entity -> Entity -> Entity ;`
  so that the corresponding branch of the tree for the question *In which team did **Dave Bing and Ron Reed** started their basketball career?* looks like:
  `EntityAndEntity (twoWordEnt "Dave" "Bing") (twoWordEnt "Ron" !"Reed"),`

- a property of an entity:
  PropOfEnt_to_Entity : Property -> Entity -> Entity
  *What is the alma mater of **the successor of F. A. Little, Jr.**?*
  PropOfEnt_to_Entity successor_0 (fourWordEnt "F." "A."
  "Little," "Jr.")
- a simple entity followed by some property:
  EntProp_to_Entity : Entity -> Property -> Entity
  *Name **Ivanpah Solar power facility owner**.*
  EntProp_to_Entity (fourWordEnt "Ivanpah" "Solar" "power"
  "facility") owner_0.
- a class with a relative clause:
  Class_to_Ent : ClassChunk -> RelCl -> Entity
  *What are the notable works of **the person who produced Queer as Folk**?*
  Class_to_Ent (Class_Chunk Person_Class) (WhoVP_relcl
  (VPSlash_to_VPChunk (V2_to_VPSlash produce_V2) (threeWordEnt
  "Queer" "as" "Folk")))

3.8   Question

A question's type is determined by the topmost rule in its parse, constructing the category Q. Currently the GQA grammar has 53 question-building rules, such as:

WhoVP : VPChunk -> Q ;
Who [developed Google Videos]?

WhereIsXDone : Entity -> V2 -> Q ;
Where was [WiZeefa] [founded]?

WhatPropOfEntIsPropOfEnt : Property -> Entity -> Property ->
Entity -> Q ;
Which [home town] of [Pavel Moroz] is the [death location] of [Yakov Estrin]?

WhatIsClassRelCl : ClassChunk -> RelCl -> Q ;
What are the [schools] [whose city is Reading, Berkshire]?

WhatClassVPAndVP : ClassChunk -> VPChunk -> VPChunk -> Q ;
Which [royalty] [was married to ptolemy XIII Theos Philopator] and [had mother named Cleopatra V]?

Note that structural words (*who, that, is, the, an,* etc.) are included only in the linearisation definitions of the rules and are not "visible" in parse trees, which is one of the main distinctions of our conceptual grammar from the "linguistic" RGL grammar. Certain words are made optional, for example, *different* in *Count the different types of Flatbread.*

One could also notice that the GQA grammar involves chunks, i.e. grammatically independent elements, which leave a possibility for ungrammatical constructions, e.g. *Which are the **television show** which have been created by Donald Wilson?* This example would have the same parse tree as a question *Which is an television shows which was created by Donald Wilson?*

4   Parse Analysis and Answer Extraction

4.1   Finding the Best Parse

The GQA grammar is ambiguous, i.e. a question can have several parses. One reason is that the grammar can consider a row of up to 10 tokens as an entity. For example, the phrase *the designer of REP Parasol* can be interpreted as

```
PropOfEnt_to_Entity designer_O (twoWordEnt "REP" "Parasol"),
```

meaning that the system detected the entity *REP Parasol* and its property *designer*. At the same time, the system can output an alternative reading

```
fiveWordEnt "the" "designer" "of" "REP" "Parasol",
```

which would not allow us to resolve the reference correctly. Thus, in choosing the parse that is further passed for interpretation we prioritise the one that has the least quoted strings in it. If this selection still gives us two or more parses with an equal number of "unresolved" strings (that is, components of entity names), we apply other prioritisation rules, e.g. preference of idiomatic constructions, class names over property names (which can sometimes coincide), and ontology property names over infobox property names (ontology properties are used in the *Functions* database as keys).

4.2   Unfolding the Parse and Querying the Index

The analysis of a selected parse starts from detecting the top function, determining the question type, and its arguments. Each argument is analysed separately, going from the outer function to the innermost one. For example, the best parse of the question *What are the awards won by the producer of Elizabeth: The Golden Age?* looks like:

```
WhatIsX (Class_to_Ent (Class_Chunk Award_Class) (WhoVP_relcl
(VPSlash_to_VPChunk (BeDoneBy_VPslash (DoneBy_PP win_V2)
 (PropOfEnt_to_Entity producer_O (fourWordEnt "Elizabeth:" "The"
"Golden" "Age")))))).
```

The innermost components of the parse are the property *producer* and entity *Elizabeth: The Golden Age*. The system looks for the entity's link, then checks that it contains the *producer* property. If not, the system tries to check if the entity is ambiguous and look through other readings. If the property is found, its value (another entity link(s)) is passed to the next level, where we look for the property *award* or *awards*. Thus, we "unfold" the parse layer by layer, starting from the inner components and making queries to DBpedia, so that we can be sure that the search is conducted in the right direction.

4.3   Spell Checking

GQA employs a string matching algorithm to tackle spelling errors in input questions. If the system cannot find a DBpedia page for a phrase that it considers as an entity name, it uses closest match by an approximate string matching method[3] to find the most likely entity page. In future we are planning to improve entity resolution by employing entity linking systems, such as DBpediaSpotlight [7] and FOX [14] (for a survey on entity linking and resolution, see [20]).

If the answer is still not found, the system tries to match question words against the GQA grammar's vocabulary (about 2500 words). For question words not found in the vocabulary, the system takes the one having the closest match to a vocabulary word by the string matching method, and corrects it to that match.

5   Testing

The GQA system was tested on the Scalable Question Answering Challenge (SQA) over Linked Data training set [15]. Participants of the SQA challenge are provided with an RDF dataset (DBpedia 2016-04) and a large training set (5000 questions) of questions with corresponding correct answers and SPARQL queries retrieving those answers. The evaluation of performance is based both on the number of correct answers and time needed.

The GQA system's performance results are presented in Table 1.

Table 1. Performance of GQA over the SQA training test set

Number of questions	Questions answered	Questions answered correctly	Precision
5000	1384 (27.68%)	1327 (26.54%)	95.88%

Table 2. The correlation of simple and complex questions answered by GQA

Simple questions answered	Complex questions answered	Simple questions answered correctly	Complex questions answered correctly	Simple questions precision	Complex questions precision
520	864	484	843	93.08%	97.57%

The majority of questions in the SQA task (about 68%) can be considered as complex questions, i.e. those questions whose SPARQL answers involve more

---

[3] *Get_close_matches* in Python's difflib finds matches, *ratio* computes distances.

than one RDF triple. The GQA system managed to correctly answer 484 simple questions (36.47% of the number of correctly answered questions) and 843 complex questions (63.53% of the number of correctly answered questions) (Table 2). Complex questions sometimes can seem confusing even for a human:

- *How many other architect are there of the historic places whose architect is also Stanford White?*
- *How many TV shows were made by someone who was associated with Lewis Hamilton?*
- *Which siler medalist of the Tennis at the 2012 Summer Olympics Men's singles was also the flagbearer of the Switzerland at the 2008 Summer Olympics?* (*sic*)

The reason for the relatively modest score for simple question answering is the system's current inability to analyse previously unseen formulations. However, if the system managed to output some answer, it was almost always correct, thus gaining the precision of 95.88%. The rare mistakes in the output queries were mostly related to incorrect entity resolution.

Further analysis of the output revealed that the GQA system often has a stricter approach to answer selection than the one used in the ground truth answer SPARQL queries. This is mostly related to class checking: whenever the GQA system detects some class in the parse, it outputs only those entities that belong to this class, whereas it is not always true in the SQA training set. For example, the question *Which politician is currently ruling over Rishkiesh?* (*sic*) has the answer http://dbpedia.org/resource/Indian_National_Congress, which does not belong to the *Politician* class. Another example is the question *What are the beverages whose origin is England?* having the SPARQL query

```
SELECT DISTINCT ?uri WHERE {?uri <http://dbpedia.org/ontology/
origin> <http://dbpedia.org/resource/England>},
```

which outputs everything that originated in England, not only beverages. We did not check all 5000 questions of the training set, but at least found 3 similar mistakes in the first 100 questions.

Correctly answered questions took 0.2–1 s in general. Answering questions with highly ambiguous elements could take up to 45 s. About 10 s was needed each time when the system failed to produce an answer.

On the whole, testing proved our assumption that the main shortcoming of the GQA system is its focus on the controlled language, which could be improved by some methods that tolerate broader variation of question wording, e.g. lexical matching of DBpedia properties and question tokens excluding entity names, which was implemented in [21]. At the same time, the grammatical approach can be quite reliable in defining the structure of a question, especially if it is syntactically complex.

6   Related Work

Numerous attempts have been made to develop question-answering systems involving various KBs and techniques: learning subgraph embeddings over Freebase, WikiAnswers, ClueWeb and WebQuestions [4], viewing a KB as a collection of topic graphs [29], learning question paraphrases over ReVerb [12], using domain-specific lexicon and grammar [27], automated utterance-query template generation [1], using multi-column convolutional neural networks [10], etc. A number of efforts were made to develop semantic parsing-based QA systems [3, 11].

Our idea of employing GF in question answering was inspired by the systems squall2sparql [13] and CANaLI [18], which got the highest scores in answering English language questions in the multilingual tasks of QALD-3 and QALD-6 [19] respectively. These systems accept questions formulated in a controlled natural language, which reduces ambiguity and increases the accuracy of answers. However, this approach requires manual reformulation of questions into the format acceptable for the systems. We develop a more extensive controlled language, which tolerates various question formulations, making the question answering process totally automatic.

GF was successfully applied in the QALD-4 question answering challenge for Biomedical interlinked data [17]. Van Grondelle and Unger [28] exploited GF in their paradigm for conceptually scoped language technology, reusing technology components and lexical resources on the web and facilitating their low impact adoption. GF was also used to convert natural language questions to SPARQL in the English [6] and Romanian cultural heritage domain [16].

7   Conclusion

GQA is an attempt to use the grammatical approach in resolving mostly complex, syntactically rich questions and converting them into conceptual parses, which then can be mapped on to DBpedia components. The approach involving controlled natural language, on the one hand, increases the probability of understanding questions accurately and retrieving correct answers. On the other hand, one would never foresee all possible formulations of natural language questions, even having large training sets at one's disposal. Nevertheless, our approach requires no manual reformulation of individual questions, unlike some other controlled language systems. In the future we plan to develop additional methods to provide better semantic analysis. Moreover, the system needs more sophisticated entity linking methods, spell checking techniques, and the implementation needs to be improved to utilise parallel computing in a multi-server, multi-processor environment. In addition, we are planning to exploit GF's multilinguality facilities and extend the system by adding new languages.

References

1. Abujabal, A., Yahya, M., Riedewald, M., Weikum, G.: Automated template generation for question answering over knowledge graphs. In: WWW 2017, pp. 1191–1200. New York (2017)
2. Lucene, A.: http://lucene.apache.org/. Last Accessed 15 June 2018
3. Berant, J., Chou, A., Frostig, R., Liang, P.: Semantic parsing on freebase from question-answer pairs. EMNLP **2013**, 1533–1544 (2013)
4. Bordes, A., Chopra, S., Weston, J.: Question answering with subgraph embeddings. In: Computer Science, pp. 615–620 (2014)
5. Bringert, B., Hallgren, T., Ranta, A.: GF Resource Grammar Library: Synopsis, http://www.grammaticalframework.org/lib/doc/synopsis.html. Last Accessed 15 June 2018
6. Damova, M., Dannélls, D., Enache, R., Mateva, M., Ranta, A.: Multilingual natural language interaction with semantic web knowledge bases and linked open data. In: Buitelaar, P., Cimiano, P. (eds.) Towards the Multilingual Semantic Web, pp. 211–226. Springer, Heidelberg (2014). https://doi.org/10.1007/978-3-662-43585-4_13
7. DBpediaSpotlight, https://www.dbpedia-spotlight.org/. Last Accessed 15 June 2018
8. DBpedia version 2016–04, http://wiki.dbpedia.org/dbpedia-version-2016-04. Last Accessed 15 June 2018
9. Diefenbach, D., Singh, K., Maret, P.: WDAqua-core0: a question answering component for the research community. In: Dragoni, M., Solanki, M., Blomqvist, E. (eds.) SemWebEval 2017. CCIS, vol. 769, pp. 84–89. Springer, Cham (2017). https://doi.org/10.1007/978-3-319-69146-6_8
10. Dong, L., Wei, F., Zhou, M., Xu, K.: Question answering over Freebase with multi-column convolutional neural networks. In: Proceedings of the 53rd Annual Meeting of the Association for Computational Linguistics and the 7th International Joint Conference on Natural Language Processing, pp. 260–269 (2015)
11. Dubey, M., Dasgupta, S., Sharma, A., Höffner, K., Lehmann, J.: AskNow: a framework for natural language query formalization in SPARQL. In: The Semantic Web. Latest Advances and New Domains, pp. 300–316 (2016)
12. Fader, A., Zettlemoyer, L., Etzioni, O. Paraphrase-driven learning for open question answering. In: Proceedings of the 51st Annual Meeting of the Association for Computational Linguistics, pp. 1608–1618 (2013)
13. Ferré, S.: squall2sparql: a translator from controlled english to full SPARQL 1.1. In: Working Notes of Multilingual Question Answering over Linked Data (QALD-3). Valencia, Spain (2013)
14. FOX: Federated Knowledge Extraction Framework, http://fox-demo.aksw.org/. Last Accessed 15 June 2018
15. LC-QuAD: A Corpus for Complex Question Answering over Knowledge Graphs, ISWC 2017, https://project-hobbit.eu/challenges/sqa-challenge-eswc-2018/. Last Accessed 15 June 2018
16. Marginean, A., Groza, A., Slavescu, R.R., Letia, I.A.: Romanian2SPARQL: A grammatical framework approach for querying linked Data in Romanian language. In: International Conference on Development and Application Systems (2014)
17. Marginean, A.: Question answering over biomedical linked data with grammatical framework. Semant. Web 8(4), 565–580 (2017)
18. Mazzeo, G. M., Zaniolo, C.: Question answering on RDF KBs using controlled natural language and semantic autocompletion. In: Semantic Web 1, pp. 1–5. IOS Press (2016)

19. Question Answering over Linked Data (QALD), https://qald.sebastianwalter.org/. Last Accessed 15 June 2018
20. Christophides, V., Efthymiou, V., Stefanidis, K.: Entity resolution in the web of data. Synthesis Lectures on the Semantic Web: Theory and Technology. Morgan & Claypool Publishers (2015)
21. Radoev, N., Tremblay, M., Gagnon, M., Zouaq, A.: AMAL: answering french natural language questions using DBpedia. In: Dragoni, M., Solanki, M., Blomqvist, E. (eds.) SemWebEval 2017. CCIS, vol. 769, pp. 90–105. Springer, Cham (2017). https://doi.org/10.1007/978-3-319-69146-6_9
22. Ranta, A.: Grammatical Framework Tutorial, http://www.grammaticalframework. org/doc/tutorial/gf-tutorial.html. Last Accessed 15 June 2018
23. Ranta, A.: Grammatical Framework: Programming with Multilingual Grammars. CSLI Publications, Stanford (2011)
24. Roy, S.B., Stefanidis, K., Koutrika, G., Lakshmanan, L.V.S., Riedewald, M.: Report on the third international workshop on exploratory search in databases and the web (ExploreDB 2016). SIGMOD Record **45**(3), 35–38 (2016)
25. Stefanidis, K., Fundulaki, I.: Keyword search on RDF graphs: it is more than just searching for keywords. In: Gandon, F., Guéret, C., Villata, S., Breslin, J., Faron-Zucker, C., Zimmermann, A. (eds.) ESWC 2015. LNCS, vol. 9341, pp. 144–148. Springer, Cham (2015). https://doi.org/10.1007/978-3-319-25639-9_28
26. SPARQL 1.1 Overview - W3C, http://www.w3.org/TR/sparql11-overview/. Last Accessed 15 June 2018
27. Unger, C., Cimiano, P.: Pythia: compositional meaning construction for ontology-based question answering on the semantic web. In: Muñoz, R., Montoyo, A., Métais, E. (eds.) NLDB 2011. LNCS, vol. 6716, pp. 153–160. Springer, Heidelberg (2011). https://doi.org/10.1007/978-3-642-22327-3_15
28. van Grondelle, J., Unger, C.: A three-dimensional paradigm for conceptually scoped language technology. In: Buitelaar, P., Cimiano, P. (eds.) Towards the Multilingual Semantic Web, pp. 67–82. Springer, Heidelberg (2014). https://doi.org/ 10.1007/978-3-662-43585-4_5
29. Yao, X., Van Durme, B.: Information extraction over structured data: question answering with Freebase. In: Proceedings of the 52nd Annual Meeting of the Association for Computational Linguistics, pp. 956–966 (2014)

A Language Adaptive Method for
Question Answering on French and English

Nikolay Radoev[1]([⊠]), Amal Zouaq[1,2], Mathieu Tremblay[1], and Michel Gagnon[1]

[1] Département de génie informatique et génie logiciel, Polytechnique Montréal,
Montreal, Canada
`{nikolay.radoev,mathieu-4.tremblay,michel.gagnon}@polymtl.ca`
[2] School of Electrical Engineering and Computer Science, University of Ottawa,
Ottawa, Canada
`azouaq@uottawa.ca`

Abstract. The LAMA (Language Adaptive Method for question Answering) system focuses on answering natural language questions using an RDF knowledge base within a reasonable time. Originally designed to process queries written in French, the system has been redesigned to also function on the English language. Overall, we propose a set of lexico-syntactic patterns for entity and property extraction to create a semantic representation of natural language requests. This semantic representation is then used to generate SPARQL queries able to answer users' requests. The paper also describes a method for decomposing complex queries into a series of simpler queries. The use of preprocessed data and parallelization methods helps improve individual answer times.

Keywords: Question answering · Linked Data · RDF · DBpedia
Scalability · SPARQL

1 Introduction

An important aspect of using the Semantic Web to its full potential is providing typical users with an easier way to access the growing amount of structured data available in different databases [1]. General knowledge bases (KBs), such as DBpedia [2], contain information extracted from Wikipedia, while many specialized knowledge bases have curated domain-specific knowledge, such as Dailymed [3]. However, given their reliance on SPARQL, these knowledge bases are difficult to use for the average user. Moreover, querying these KBs without knowledge of their underlying structure is a very complex task. Developing an intuitive interface to allow natural language queries is still an open problem [4].

The Scalable Question Answering over Linked Data Challenge (SQA) requires finding a way to transform a question into a query over a knowledge base, DBpedia being the one used in this particular challenge. Questions range from simple ones, which target a single fact about a single entity, to complex

© Springer Nature Switzerland AG 2018
D. Buscaldi et al. (Eds.): SemWebEval 2018, CCIS 927, pp. 98–113, 2018.
https://doi.org/10.1007/978-3-030-00072-1_9

questions that involve many entities and require some kind of additional computation or inference to provide an answer. The training data provided in the challenge consists of 5000 natural language questions, for which one or more answers must be extracted from DBpedia.

In addition to being able to answer questions, systems must also be able to give those answers in a reasonable time and be able to handle a large load of queries in a given time. Assuming that normally, a QA system is not ran locally by the client, network delay in the user's requests must also be taken into account. If the QA system itself acts as a client to external services, the network delay becomes an important factor in the system's final performance.

We present LAMA (Language Adaptive Method for question Answering), an extension to AMAL [5], a system that was originally created for the QALD2017 [1] Challenge at the ESWC2017 [6] conference. The first version of our system was built specifically to handle questions in French. In LAMA, we extend this capability to the English language and handle more complex questions. Given that the AMAL system was based on the French syntactic structures, additional work was done to reduce the importance of language-specific structures. As an example, French often uses a *Subject-Object-Verb* structure (e.g. *Il le mange*, (*He it eats*) in English), compared to English where the *Subject-Verb-Object* structure is used instead (e.g. *He eats the apple*). In addition, the LAMA system does not enforce a controlled natural language [7] for the input queries but handles the questions "as-is".

Our approach aims at minimizing the number of manual features and ad hoc heuristics required to process questions. We also aim to optimize the preprocessing time for each question to improve scalability for a large number of questions or to deal with complex questions which may be costly in terms of computational resources. Finally, we propose a method to decompose complex questions into a series of smaller, simpler queries that can be more easily processed. This paper is a more detailed version of the one initially submitted [8] to the SQA2018 challenge.

While DBpedia is mentioned throughout the article, the LAMA system is not completely specific to the DBpedia knowledge base. Other KBs, such as WikiData [9], can be used to answer questions as long as some conditions are respected, notably the availability of the property configurations files Sect. 3. For readability purposes, we have abbreviated certain URIs by using the prefixes detailed in Table 1. For example, http://dbpedia.org/ontology/spouse becomes *dbo:spouse*. The *dbo:* prefix is used to represent classes (aka concepts) from the DBpedia ontology while the *dbr:* prefix is used to identify resources from this KB.

2   Related Work

Question Answering over linked data is a problem that has already been explored by different studies in the past. Generally, most QA systems follow a similar approach to produce answers: the user's question is taken as input, parsed to

Table 1. DBpedia prefixes

dbo	http://dbpedia.org/ontology/
dbr	http://dbpedia.org/resource/
dbp	http://dbpedia.org/property/

extract the various relations between the entities and then a query (most often written in SPARQL) is generated and submitted to one or more KBs.

Generally, all systems try to build a semantic representation of the question that relies mainly on the identification of entities and their properties in the user's questions. Xser [10] and WDAqua-core1 [11] rely on string matching and generate all possible interpretations for every word in the question, i.e. each word can be considered a potential entity or property. QAnswer [12] uses the Stanford CoreNLP parser to annotate questions and employs a mix of string matching with both text fragments and synonyms extracted from WordNet and the information given by the POS tags and lemma provided by the parser. This method has the added benefit of not relying on a perfect matching, making it more resistant to spelling mistakes.

While the general approach is similar in all the state of the art, the way the questions are parsed and what information is extracted vary among systems. Most systems [10,13,14] rely on semantic structures in order to find the answer to a given question. Xser defines a semantic structure as a structure that can be instantiated in the queried data [10]. For instance, SemGraphQA [13] generates direct acyclic graphs that represent all possible and coherent interpretations of the query and only keeps graphs that can be found in DBpedia's graph representation. WDAqua-core1 has a similar approach to [13] that ignores the syntax and focuses on the semantics of the words in a question by extracting as many entities as possible and trying to find relationships between them by exploring the RDF graph of each entity. Relying on partial semantics, where only a part of the question is given a semantic representation, appears to be a reliable approach towards QA given its wide use [10,11,13]. Finally, all of the systems observed used DBpedia or Wikidata as their KB backend, making the use of SPARQL queries ubiquitous. WDAqua-core1 generates multiple SPARQL queries and ranks them based on multiple criteria, mainly the number of words in the question that are linked to resources in the query and how similar the label is to the corresponding resource. In contrast, QAnswer generates a single SPARQL query at the end of its entity-relationship extraction phase.

The approach proposed with LAMA is also based on the identification of entities and properties. However, unlike Xser [10] and others, our entity and property extraction do not rely only on string matching but also on lexico-syntactic patterns, detailed in Sects. 3.2 and 3.3. Additionally, our SPARQL query generator creates multiple queries in order to explore different possible interpretations of a question but also has some additional pre-processing to limit unnecessary query generation, unlike WDAqua-core1.

3   System Overview

As previously mentioned, LAMA is based on the AMAL system developed for the QALD-7 challenge and thus keeps the same core architecture. The system is developed using a modular approach to separate application logic in different subsystems. Each subsystem can be developed, modified or enabled independently. For example, the *translation* module can be toggled on or off based on the initial language of the query: in the original system, parts of the French requests were translated into English in order to retrieve information from the English version of DBpedia. While we have attempted to reduce the reliance on language-specific rules, all main modules of the LAMA system can be extended and adapted to include such rules if the developers deem it appropriate.

Originally, AMAL's main focus was to interpret and answer *Simple Questions*, which are questions that concern (i) only one single entity and a property of this entity, e.g. *Who is the mayor of Montreal?* where *Montreal* is the **Entity** and *dbo:mayorOf* is the **Property** or (ii) a single relationship between two entities (e.g *is Valerie Plante the mayor of Montreal?*). The LAMA system now also handles some *Complex Questions* by converting them into a set of *Simple Questions* and processing those smaller parts. The architecture of our system involves a multi-step pipeline: a *Question Type Classifier*, an *Entity Extractor*, a *Property Extractor* and a *SPARQL Query Generator* that generates the final SPARQL query that is used to retrieve the answer(s) from DBpedia (Fig. 1).

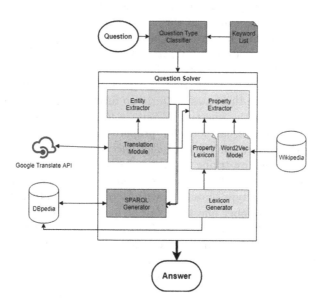

Fig. 1. System overview

The first step is to determine the type of the question and the nature of the expected answer. Once the system knows the question type, the query is sent to a specific solver that has been designed for this type of question. For instance, a question such as *Is GIMP written in GTK+?* will be sent to the *Boolean* answerer, and *Who died due to Morphine?* is handled by the *Resource* question solver. Every question solver makes use of one or more submodules that function as *extractors*. There are two main extractors: an *entity extractor* and a *property extractor*. Once we obtain a semantic representation of the natural language question using knowledge base entities and properties, we generate one or more SPARQL queries based on these entities and properties. Given that multiple valid entities and/or properties can be found in a single query, all possible SPARQL queries based on these valid combinations are generated and saved in a queue.

The following sections detail the main steps of our system's pipeline: the Question Type Classifier, the Entity Extractor, the Property Extractor and the SPARQL Query Generator and their general implementation. As previously mentioned, each module can be extended and modified separately.

3.1   Question Type Classifier

Currently, questions are grouped into the following types: *Boolean, Date, Number* and *Resource. Resource* is the most generic type and designates questions for which the expected answer is one or more URIs. It is used as the default type for questions that don't fit any of the other categories. Boolean questions operate on the closed world assumption and always return a *TRUE* or *FALSE* answer, with *FALSE* being returned in the case where the answer is unknown. Date questions refer to dates in a standard *YYYY-MM-DD* format. The question type is determined through pattern matching with manually extracted patterns (roughly 5 patterns per question type) from the QALD6 and QALD7 training sets, totalling more than 400 questions. The different patterns and an example for each are given in more detail in Table 2. Questions about location are classified as *Resource* but have their own patterns to further help with classification. For example, classifying the question *Where is the Westminster Palace?* as a location question type, we can look for properties such as *dbo:location* or at least properties that are in the range of *dbo:Place*.

Classification is done by trying to match the questions to different types in the following order: Boolean, Date, Number, and Resource. This order is due to the relative complexity of type matching, with Boolean and Date questions being easier to detect while Number questions can be more complicated, i.e. requiring counting the number of answers in a set (e.g. *How many languages are spoken in Canada?*) or calculating a timespan (e.g. *How long did World War II last?*). In the current version of the system, multiple types are not supported and only the first detected type is considered. The system does however support subtypes, more specifically, the *Aggregation* subtype that applies to questions that require counting or ordering (ascending or descending) of results. Several examples are provided in Table 3.

This pattern-based method has the advantage of being easily transferable between languages in the context of a multilingual system, as patterns can be easily expressed in different languages. In fact, after extracting patterns for both

Table 2. Keywords for question classification

Contains	Example
	Boolean question
Does	Does Toronto have a hockey team?
Do	Do hockey games half halftime?
Did	Did Ovechkin play for Montreal?
Was	Was Los Angeles the 2014 Stanley cup winner?
Is	Is Corey Crawford Blackhawks' goalie?
	Date question
When	When was the operation overlord planned?
What time	What time is the NHL draft?
What date	What date is mother's day?
Give me the date	Give me the date of the FIFA 2018 final?
Birthdate	What is the birthdate of Wayne Gretzky?
	Location question
Where	Where is the Westminster palace?
In what country/city	In what country is the Eiffel tower?
(Birth)place	What is Barack Obama's birthplace?
In which	In which state is Washington DC in?
The location of	What was the location of Desert Storm?

Table 3. Number and aggregation question classification

Contains	Example
How many	How many countries were involved in World War 2?
Count	Count the TV shows whose network company is based in New York?
List the [..]	List the categories of the animal kingdom.
Sum the things	Sum the things that are produced by US companies?
the most/the least	What is the most important worship place of Bogumilus?
first/last	What was the last movie with Alec Guinness?
[ADJ]-est (biggest, oldest, tallest, etc.)	Name the biggest city where 1951 Asian Games are hosted? Who is the tallest player in the NBA?

French and English, the system was able to accurately predict the question type in more than 92% of the QALD7 training set.

3.2 Entity Extraction

As previously mentioned, the original AMAL project focused on *Simple Questions*, limited to at most one entity and one property. For example, a simple question is *Who is the father of Barack Obama?*, where 'Barack Obama' is the entity and 'father of' is the property. To extract the entity from the question, we first try to match the words (or a combination of these words) to existing DBpedia entities. Given that noun groups can contain many more elements than just the entity, such as adjectives or determinants, we start by removing the question keywords and consider the remaining string as the longest possible entity. We then generate all possible substrings by recursively tokenizing the output. All generated combinations are then appended to the *dbr:* prefix (see Table 1), e.g. *dbr:Barack_Obama* and are ran against the DBpedia database to find as many valid (i.e. that exist in DBpedia) candidate entities as possible. For example, *Who is the queen of England?* generates the following substrings after removing the question indicator (*Who is*, a *Resource* question indicator): *queen of England, queen, England, queen of, of England.* Out of those, only the first 3 are kept as valid entities since *queen of* and *of England* are not DBpedia entities. The selected entities are then sorted by length.

In order to increase the set of considered entities, we add variations of the identified nouns using normalization and stemming. For example, the noun *queen* can be transformed to extract the entities *Queens* or *Queen*. If no entities are found using this method, we use the DBpedia Spotlight [15] tool. Spotlight is a tool for entity detection in texts and its performance is limited when applied to single phrases so it is only used as a backup method.

Once all the possible candidate entities have been extracted, we use multiple criteria, such as the length of the entity's string, the number of modifications needed to find it in DBpedia and whether we had to translate it, to determine their likelihood of being the right entity. For example, every modification through normalization and stemming reduces the likelihood of the entity to be chosen. In our previous example, since *Queens* requires 2 modifications (pluralization and capitalization), it is less likely to be selected as the correct entity by the entity extractor. The formula we currently use to combine these factors is the following, where e is the entity string:

$$length(e) - T \times \frac{length(e)}{2} + 2 \times U \times nsp(e) - P$$

where:
   $nsp(e)$ is the number of spaces in e
   P is the number of characters added/removed to add/remove pluralization
   $T = 1$ if e has been translated, 0 otherwise
   $U = 1$ if e has not been capitalized, 0 otherwise

According to the formula, the score is penalized by half the length of the entity string if we used a translation. This only applies for languages other than English since the translation step is entirely skipped when analyzing an English query. Also, the more words it contains (a word is delimited by the use of spaces) the higher the score will be, if it has not been capitalized. For example, *In which city is the Palace of Westminster in?* we can extract *Palace of Westminster* and

Westminster. However, using the above formula, *Westminster* has a score of 11 while *Palace of Westminster* has a score of 25. Therefore, we consider it more likely that the correct entity is *Palace of Westminster*.

During the entity extraction phase, there is very often the issue of ambiguity when deciding which resource in the knowledge base we are looking for. For example, *Victoria* can refer to: *Person, Ship, City, Lake, etc.* To determine which resource to use to answer the query, LAMA has a multi-step approach which relies not only on candidate entities but also on properties, as explained in Sects. 3.2 and 3.3:

1. For every candidate entity that is a disambiguation page (e.g. *dbr:Victora, dbr:Victoria_ (ship), dbr:Queen_ Victoria_ (ship), dbr:Victoria_ (name) etc.*) for the word *Victoria*, we extract possible disambiguation entities following the disambiguation link provided in DBpedia and store them in a vector V where the first element is the original entity (i.e. *dbr:Victoria*).
2. For each entity in the vector V, we generate a SPARQL query that involves the entity and the properties extracted by the *Property Extractor* (see Sect. 3.3). We keep only the entities whose queries return non-empty answers for an *ASK* SPARQL query looking for an existing link between the entity and the property.
3. In the case of multiple entities having the same property in their DBpedia description, we find the intersection between the entity's label and the original query's string representations (character per character) and compute its length (See the example below). Results are then sorted by a distance factor (from 0 to 1) based on the ratio of the intersection's value and the entity length using the following representation:

$$distance\ vector = \frac{|intersection\ set|}{entity's\ label\ length}$$

This can be best illustrated with an example using the following question taken from the SQA training set: *Ship Victoria belongs to whom?* The property in the query is: *dbo:owner* (representing the expression *belongs to*) but there are 2 candidate entities: *dbr:Ship* and *dbr:Victoria* and as already mentioned, *dbr:Victoria* has several possible disambiguations (more than a 100). Most of those can be eliminated as they are not linked to any other entity by the *dbo:owner* property. However, some resources do have the property and can be valid possibilities, most notably: *Victoria_ (ship)* and *Queen_ Victoria_ (ship)* where both entities have a *dbo:owner* and are both *ships*. In this case, calculating the intersection's length between those entities and the original question's string, we get an intersection length of **12** (i.e. there are 12 letters shared between the two strings) for *Victoria_ (ship)* and **13** for *Queen_ Victoria_ (ship)*. However, the distance factor is **12/12** and **13/17** respectively. Using those metrics, we return a list of ranked entities where *Victoria_ (ship)* is the top entity.

3.3   Property Extraction

Besides the entity extraction, another step for the request's semantic representation involves the extraction of properties from the original question to generate SPARQL queries. First, the potential entities tagged by the *Entity Extractor* are removed from the query and the remaining tokens are parsed for potential properties. The process is actually run in parallel with the entity extraction since some parts of it depend on the existence of entity-property relations. This also helps with the processing time, since we can do both extractions at the same time.

To increase performance, we have automatically extracted more than 20000 DBpedia properties along with their labels, thus building a property lexicon. The information is stored in a hash table, allowing for a fast ($O(1)$) access to a property URI based on its label. Without this lexicon, verifying if a possible label is related to a valid DBpedia property would require sending a network query to DBPedia, which incurs additional time and computational investment. While much of the *label-URI* pairs can be considered noise, e.g. *dbp:125TotalPoints*, some can be quite useful for question answering tasks, e.g. ["author": dbo:author], ["spouse": dbo:spouse], ["birth place": dbo:birthDate] etc. In the case of properties that exist in both **dbo** and **dbp** prefixes Table 1, both properties are added to the *label-URI* pair with the **dbo** URI having priority, e.g. *"parent"* is a label with both *dbp:* and *dbo:* prefixes, but most entities use the *dbo:* prefix. Querying for the properties and their labels, with the additional option to only get labels in a specific language (English in this case) can be done using this SPARQL query to DBpedia:

```
SELECT ?pred ?label WHERE {
    ?pred a rdf:Property .
    ?pred rdfs:label ?label
FILTER (
    LANG(?label)="" || LANGMATCHES(LANG(?label), "en"))
}
```

This allows us to easily find existing properties based on their label and its presence as a text fragment in the query. If the entities identified in the natural language request do not have any of the potential properties in their description, they are simply discarded.

The original AMAL system relied on manually annotated mappings between properties and entities to tackle ambiguities in property selection. This initial reliance on manual bindings has been reduced in LAMA by adding some basic rules for valid entity-property relations. For example, *the tallest* can be ambiguous given that it can both mean *height* or *elevation*, with the first meaning used in a relation with the DBpedia entity type *Person*, and the second used with entities that are instances of *Natural Place*. A valid rule for *the tallest* would thus be: *the tallest: {height: dbo:Person}, {elevation: dbo:NaturalPlace}*

Having such type-based rules can improve performance and speed by exploring only properties that are part of a valid rule. While those rules are still manually introduced in the system, generalizing them to the most generic type eliminates the need to add a rule for each instance of the targeted class.

Unlike entity detection, where entities are often detectable using simple string matching with the use of some lemmatization techniques, properties are rarely available in their literal form in a question. For example, the question: *Who played Agent Smith in Matrix?* can be answered by using the DBpedia property *dbo:portrayer*, a word not present in the original query.

To address this issue, LAMA uses a Word2Vec [16] model to expand the possible properties that can be extracted from a natural language query. The Word2Vec model is trained on data from Wikipedia in both French and English. The trained model provides a list of words that are similar to a given word along with their cosine distance. Currently, the Word2Vec model only applies to single words so complex expressions are only handled by the lexicon. If a word is selected as a possible property label, but it does not match one in the property lexicon, LAMA explores the similarity vector of the word, looking for a label that is contained in the lexicon. Only words with a cosine distance above 0.5 are kept to minimize false positives.

As an example, using the question *Is Michelle Obama the wife of Barack Obama?*, the word *wife* is identified as the label of a potential property but is not a label present in the property lexicon. Using the Word2Vec similarity vector for *wife*, we find *spouse* with a cosine distance of 0.723, a word that is also an existing label (mapped to *dbo:spouse*) in the lexicon. Using this, the following SPARQL query can be generated to find the answer to the question:

```
ASK WHERE {
    <dbr:Barack_Obama dbo:spouse dbr:Michelle_Obama.>
}
```

3.4   SPARQL Query Generation

The last step is building and sending SPARQL queries to DBpedia. The system uses the standard DBpedia endpoint: http://dbpedia.org. For now, LAMA supports basic SPARQL queries such as *ASK* (for checking for existing entities and *Boolean* questions) and *SELECT*. The *SPARQL Query Generator* supports the **ORDER BY** modifier, which allows for sorted answers or a single answer, e.g. "What is the *highest* mountain...?" or "Who was the *youngest* prime minister of Canada?". Queries exist as basic templates in which specific values are injected in the form of RDF triples. When a question has multiple possible *entity-property* combinations, the generated SPARQL queries are sent in parallel since the different combinations are independent from each other. Only queries that return a non-null answer are kept, with the exception of *ASK* queries where an empty answer is considered **FALSE**.

4   Complex Question Simplification

As previously indicated, AMAL focused on *Simple Questions* and as defined, a *Simple Question* is a query that can be represented as a single RDF triple, e.g. *Is Valerie Plante the mayor of Montreal?* is represented by:

<dbo:Valerie_Plante dbo:mayor dbo:Montreal>

However, some natural language queries can be more complex, containing multiple relationships between entities in the same sentence. In LAMA, we aim to also analyze and answer such complex questions. In fact, many of those *complex* questions can be represented as a chain of smaller, simpler queries. For example, the question and its respective RDF triple representation:

> *Who were involved in the wars where Jonathan Haskell battled?*
> <dbo:Jonathan_Haskell dbo:battle **?wars**>
> <**?wars** dbo: combatant **?answer**>

can be transformed into:

> *What battles was Jonathan Haskell in?*
> <dbo:Jonathan_Haskell dbo:battle **?wars**>
> and
> *Who were the combatants in those wars?*.
> <**?wars** dbo: combatant **?answer**>

The question is thus separated in two sub-queries with *wars* being the only shared word. Currently, deciding when to split a *complex question* into multiple *simple questions* is based on the presence of a keyword Table 4, *where* being the separation marker in this example. The split is done right before the targeted keyword and the keyword is kept in the second sub-question. The following table shows the different keywords used to detect *Complex Questions*.

Table 4. Complex question keywords

Contains	Example
Where	Who were involved in the wars where Jonathan Haskell battled?
Whose	What are the TV shows whose network is also known as the CW?
Which	Give me everything owned by networks which are lead by Steve Burke?
Also	Name the scientist whose supervisor also supervised Mary Ainsworth?
Who	List the resting place of the people who served in Norwalk Trainband

A special case is made for questions that involve the following structure: [...] **both** X **and** Y [...]. In such questions, two RDF triples are automatically generated that have the same structure, but the subject of both is **X** and **Y** respectively. For example, the question:

Which university was attended by both Richard H. Immerman and Franklin W. Olin?

generates the following RDF triples:

<dbo:Richard_H._Immerman dbo:education **?university**>
<dbo:Franklin_W._Olin   dbo:education **?university**>

By having a set of *Simple Questions*, each question can be analyzed separately and with a reduced amount of tokens in each phrase, the system has a lower probability to generate erroneous entity-property combination, e.g. linking *Jonathan Haskell* and *involved*(dbo:combatant) together in our example *Who were involved in the wars where Jonathan Haskell battled?*. Currently, the system waits for all triples to be generated by each *Simple Question* and combines them in a single SPARQL query. This saves on processing and answer time, since only a single SPARQL request is sent for the whole question, instead of an individual request for each RDF triple.

5 Parallel Processing and Scalability

The original version of the AMAL system had a procedural processing cycle which could bring scalability issues. The system had to wait for every SPARQL query sent to DBpedia to finish its execution before trying a new one and there was no predefined timeout for a request. Originally, all answers were expected to be in English, even if the system was designed to handle French queries only, thus requiring an extra layer of translation for every question. Questions where multiple translations were required were the biggest problem, since the Google API does not allow for batch querying.

For the SQA challenge, we propose a redesign of the system, focusing on optimizing scalability. The following points are the main focus of the optimization techniques and target different parts of the process in order to maximize results.

– The extra step of translating French into English is no longer required, given that the questions are in English only. This saves around 15 to 20% of computation time, depending on the complexity of the question and the amount of translations necessary to process the equivalent question in French.
– Some parts of the question's processing have been moved to the client side, most notably the lookup of valid DBpedia properties, which no longer requires sending requests to DBpedia. This reduces networking time overhead and access overhead since a hashtable has an $O(1)$ complexity.
– SPARQL queries for the same question are run in parallel. Since the generated SPARQL queries are independent, running a query as soon as it is generated by the solver, without waiting to see if the previous one was successful or not, helps with the scalability of the system for multiple queries ran at the same time.

6   Evaluation

The evaluation of our system was done on the training set for the SQA2018 challenge. The base benchmark was run on the 5000 provided questions in the training set. The training set contains roughly 32% of *Simple Questions*. We have evaluated both the accuracy of the returned answers and the time to execute all queries. The evaluation was performed on a 4 core i5-4690K CPU with 8GB of RAM using a hard 5 s timeout on SPARQL queries. The tests were run on a limited environment to demonstrate that the LAMA system can be run on an average user's machine without requiring any particularly powerful hardware. The system was able to answer 2654 out of the 5000 questions, which gives us a 53.3% accuracy for the answers with a success rate of 74% for *Simple Questions* and 33.5% for *Complex Questions*. The evaluation took a total of 15603 s, giving us an average of 3.12 s per query.

When our keyword-based query separation method was applied on the SQA training set, we noted a 20 to 22% improvement in question answering. Not all returned answers sets were complete, given that the method to separate the complex queries is rather basic, but since the previous system had no way to analyze such questions, it is a notable improvement. More details on how this can be improved in future are given in Sect. 7.

While the results are not as good as the original AMAL system (74% accuracy), we have to note that the new version relies much less on case-by-case exception handling and it has transitioned from a system aimed specifically at French to a multilingual system that handles French and English questions.

Finally, the SQA challenge also involves evaluating the competing systems on a test set. We observed that, in contrast to the training set, the test set includes many errors, such as spelling errors, case errors or incorrectly tokenized questions. We expect a decrease of LAMA's performance due to these errors. To evaluate the impact of spelling mistakes, we have taken 200 questions from the SQA training set and randomly added the same type of errors (as detailed in Sect. 7) to some of the questions. This gives us an accuracy of 41% and precision of 45%, since some questions were only partially correctly answered. Based on this, it is clear that spelling has an important impact on the system's performance.

7   System Limitations and Future Work

One of the current limitations of the system is its difficulty to correctly split complex questions into multiple simple questions. Most of the errors in complex questions are due to the fact that the keyword used to split the sentence is not in fact a correct separator. For example, the sentence: *Give me the places where people who worked for HBO died in?* has been split into;

> {*Give me the places*} , {*where people*} and   {*who worked for HBO died in?*}

This ends up giving 3 sub-queries where the first one (*Give me the places*) is correct, while the second one (*where people*) does not make sense and the last one (*who worked for HBO died in?*) is interpreted by LAMA as looking for somebody who work for HBO. The final result is thus obviously incorrect.

We plan on improving those results with the processing of the *Complex Questions* using a syntactic analysis instead of relying on simple keywords. Using a syntactic parser, such as SyntaxNet, will allow for easier division of queries. Figure 2 shows the syntactic analysis of our example above. Following the POS tagging and relations, the question can be split into: *[...] people who worked for HBO [...]* and *Give me the places where X [...] died in* where X is the result of the first sub-question. This separation corresponds to the correct interpretation of the question, which can thus be answered more accurately.

Fig. 2. Syntactic parsing example

Another limitation of LAMA is that it relies on string matching for entity and property detection. This means that the system is not resistant to typographic mistakes which can easily produce false results. While working on the system, we discovered that using Google Translate can sometimes correct spelling mistakes for some words even if asked to translate from English to English. However, this method is not reliable, providing spelling correction for only around 30% of the misspelled words and can also incur a non- negligible time cost if all words are to be translated for spell checking.

Future work for the system should include a spell check module that can correct most common mistakes. The testing set for SQA2018 has quite a few spelling mistakes and over 75% of those are simple letter inversions (*What* becomes *Waht*) or single missing letter (*hve* instead of *have*). Most of those mistakes can be detected and corrected by a spell checker and thus, improve entity and property extraction in the users' questions.

8   Conclusion

In this paper we presented the LAMA system, a modification of a currently existing QA system redesigned for multilingual support and scalability performance. We have modified it by adding emphasis on scalability and reduced its dependence on ad hoc heuristics and case by case exception handling. Additional work was done for handling *Complex* queries that are defined as questions requiring a SPARQL query with multiple RDF triples. A method for splitting those complex questions in a series of simple ones was also proposed.

There is still some work to be done, most notably by improving the way complex queries are decomposed in a sequence of simpler ones. Our approach relies on a string and label matching approach for entity and property extraction in the users' questions but suffers from accuracy issues in case of spelling mistakes. In order to reduce this problem, we plan on adding a spell checking module to the system's pipeline.

Acknowledgements. This research has been partially funded through Canada NSERC Discovery Grant Program.

References

1. Qald 2017 challenge eswc 2017 hobbit. https://project-hobbit.eu/challenges/qald2017/. Accessed 29 Mar 2018
2. Auer, S., Bizer, C., Kobilarov, G., Lehmann, J., Cyganiak, R., Ives, Z.: DBpedia: a nucleus for a web of open data. In: Aberer, K. (ed.) ASWC/ISWC -2007. LNCS, vol. 4825, pp. 722–735. Springer, Heidelberg (2007). https://doi.org/10.1007/978-3-540-76298-0_52
3. National Institutes of Health et al. Daily med (2014)
4. Gupta, P.: A survey of text question answering techniques. Int. J. Comput. Appl. **53**, 1–8 (2012)
5. Radoev, N., Tremblay, M., Gagnon, M., Zouaq, A.: AMAL: answering french natural language questions using DBpedia. In: Dragoni, M., Solanki, M., Blomqvist, E. (eds.) SemWebEval 2017. CCIS, vol. 769, pp. 90–105. Springer, Cham (2017). https://doi.org/10.1007/978-3-319-69146-6_9
6. 14th eswc 2017 |. https://2017.eswc-conferences.org/. Accessed 29 Mar 2018
7. Mazzeio, G.: Answering controlled natural language questions on RDF knowledge bases (2016). https://openproceedings.org/2016/conf/edbt/paper-259.pdf
8. Radoev, N., Zouaq, A., Tremblay, M., Gagnon, M.: LAMA: a language adaptive method for question answering. In: Scalable Question Answering over Linked Data Challenge (SQA2018), Heraklion, Greece (2018)
9. Vrandečić, D., Krötzsch, M.: Wikidata: a free collaborative knowledgebase. Commun. ACM **57**(10), 78–85 (2014)
10. Xu, K., Zhang, S., Feng, Y., Zhao, D.: Answering natural language questions via phrasal semantic parsing. In: Zong, C., Nie, J.-Y. (eds.) Natural Language Processing and Chinese Computing, pp. 333–344. Springer, Berlin (2014). https://doi.org/10.1007/978-3-662-45924-9_30
11. Diefenbach, D., Singh, K., Maret, P.: Wdaqua-core1: a question answering service for rdf knowledge bases. In: Companion Proceedings of the The Web Conference 2018, WWW 2018, pp. 1087–1091, Geneva, Switzerland. International World Wide Web Conferences Steering Committee (2018)
12. Ruseti, S., Mirea, A., Rebedea, T., Trausan-Matu, S.: Qanswer - enhanced entity matching for question answering over linked data. In: CLEF (2015)
13. Beaumont, R., Grau, B., Ligozat, A.-L.: Semgraphqa@qald-5: Limsi participation at qald-5@clef. 09 2015
14. Sorokin, D., Gurevych, I.: End-to-end representation learning for question answering with weak supervision. In: Dragoni, M., Solanki, M., Blomqvist, E. (eds.) SemWebEval 2017. CCIS, vol. 769, pp. 70–83. Springer, Cham (2017). https://doi.org/10.1007/978-3-319-69146-6_7

15. Daiber, J., Jakob, M., Hokamp, C., Mendes, P.N.: Improving efficiency and accuracy in multilingual entity extraction. In: Proceedings of the 9th International Conference on Semantic Systems (I-Semantics) (2013)
16. Mikolov, T., Chen, K., Corrado, G., Dean, J.: Efficient estimation of word representations in vector space. CoRR, arXiv:1301.3781 (2013)

Semantic Sentiment Analysis Challenge

Semantic Sentiment Analysis Challenge at ESWC2018

Mauro Dragoni[1]([✉]) and Erik Cambria[2]

[1] Fondazione Bruno Kessler, Trento, Italy
`dragoni@fbk.eu`
[2] Nanyang Technological University, Singapore, Singapore
`cambria@ntu.edu.sg`

Abstract. Sentiment Analysis is a widely studied research field in both research and industry, and there are different approaches for addressing sentiment analysis related tasks. Sentiment Analysis engines implement approaches spanning from lexicon-based techniques, to machine learning, or involving syntactical rules analysis. Such systems are already evaluated in international research challenges. However, Semantic Sentiment Analysis approaches, which take into account or rely also on large semantic knowledge bases and implement Semantic Web best practices, are not under specific experimental evaluation and comparison by other international challenges. Such approaches may potentially deliver higher performance, since they are also able to analyze the implicit, semantics features associated with natural language concepts. In this paper, we present the fifth edition of the Semantic Sentiment Analysis Challenge, in which systems implementing or relying on semantic features are evaluated in a competition involving large test sets, and on different sentiment tasks. Systems merely based on syntax/word-count or just lexicon-based approaches have been excluded by the evaluation. Then, we present the results of the evaluation for each task.

1 Introduction

The development of Web 2.0 has given users important tools and opportunities to create, participate and populate blogs, review sites, web forums, social networks and online discussions. Tracking emotions and opinions on certain subjects allows identifying users' expectations, feelings, needs, reactions against particular events, political view towards certain ideas, etc. Therefore, mining, extracting and understanding opinion data from text that reside in online discussions is currently a hot topic for the research community and a key asset for industry.

The produced discussion spanned a wide range of domains and different areas such as commerce, tourism, education, health, etc. Moreover, this comes back and feeds the Web 2.0 itself thus bringing to an exponential expansion.

This explosion of activities and data brought to several opportunities that can be exploited in both research and industrial world. One of them concerns the

D. Buscaldi et al. (Eds.): SemWebEval 2018, CCIS 927, pp. 117–128, 2018.
https://doi.org/10.1007/978-3-030-00072-1_10

mining and detection of users' opinions which started back in 2003 (with the classical problem of polarity detection) and several variations have been proposed. Therefore, today there are still open challenges that have raised interest within the scientific community where new hybrid approaches are being proposed that, making use of new lexical resources, natural language processing techniques and semantic web best practices, bring substantial benefits.

Computer World[1] estimates that 70%–80% of all digital data consists of unstructured content, much of which is locked away across a variety of different data stores, locations and formats. Besides, accurately analyzing the text in an understandable manner is still far from being solved as this is extremely difficult. In fact, mining, detecting and assessing opinions and sentiments from natural language involves a deep (lexical, syntactic, semantic) understanding of most of the explicit and implicit, regular and irregular rules proper of a language.

Existing approaches are mainly focused on the identification of parts of the text where opinions and sentiments can be explicitly expressed such as polarity terms, expressions, statements that express emotions. They usually adopt purely syntactical approaches and are heavily dependent on the source language and the domain of the input text. It follows that they miss many language patterns where opinions can be expressed because this would involve a deep analysis of the semantics of a sentence. Today, several tools exist that can help understanding the semantics of a sentence. This offers an exciting research opportunity and challenge to the Semantic Web community as well. For example, sentic computing is a multi-disciplinary approach to natural language processing and understanding at the crossroads between affective computing, information extraction, and common-sense reasoning, which exploits both computer and human sciences to better interpret and process social information on the Web.

Therefore, the Semantic Sentiment Analysis Challenge looks for systems that can transform unstructured textual information to structured machine processable data in any domain by using recent advances in natural language processing, sentiment analysis and semantic web.

By relying on large semantic knowledge bases, Semantic Web best practices and techniques, and new lexical resources, semantic sentiment analysis steps away from blind use of keywords, simple statistical analysis based on syntactical rules, but rather relies on the implicit, semantics features associated with natural language concepts. Unlike purely syntactical techniques, semantic sentiment analysis approaches are able to detect sentiments that are implicitly expressed within the text, topics referred by those sentiments and are able to obtain higher performances than pure statistical methods.

The fifth edition of the Semantic Sentiment Analysis Challenge[2] followed the success, experience and best practices of the first four. It provided further stimulus and motivations for research within the Semantic Sentiment Analysis area.

---

[1] Computer World, 25 October 2004, Vol. 38, NO 43.
[2] http://www.maurodragoni.com/research/opinionmining/events/challenge-2018/.

The fifth edition of the challenge focused on further development of novel approaches for semantic sentiment analysis. Participants had to design a concept-level opinion-mining engine that exploited Linked Data and Semantic Web ontologies, such as DBPedia[3].

The authors of the competing systems showed how they employed semantics to obtain valuable information that would not be caught with traditional sentiment analysis methods. Accepted systems were based on natural language and statistical approaches with an embedded semantics module, in the core approach. As happened within the first four editions of the challenge [9–12], a few systems merely based on syntax/word-count were excluded.

The fifth challenge benefited from a Google Group that we created and named Semantic Sentiment Analysis Initiative[4] and that we opened before the Challenge proposal. Currently, the group consists of more than 200 participants and we leverage that to disseminate and promote our initiatives related to the Sentiment Analysis domain. Moreover, the fifth edition of the challenge could also benefit from a Workshop[5] we chaired at ESWC 2018 related to the same topics. Challenge had therefore an additional strength provided by the mutual support between the two events. Challenge systems were in fact invited to a Workshop dedicated session for discussing open issues and research directions showing the last technological advancements. This dual action stimulated and encouraged participants to present their work at the two events.

The remainder of the chapter is organized as follows. Section 2 discusses background work related to semantic sentiment analysis. Section 3 lists and details the tasks we have proposed in the fifth edition of the challenge as well as the annotated datasets we have used for the training, testing and evaluation phase. Section 4 shows the systems submitted by the challengers and their results are showed in Sect. 5. Finally, conclusions, considerations and our plans for the next edition of the challenge are drawn in Sect. 6.

2   Related Work

After the successes of the 2014, 2015, 2016 and 2017 editions [9–12], the ESWC conference[6] included again a challenge call with a dedicated session. The Semantic Sentiment Analysis challenge has been proposed and accepted for the fifth time on a row in the 2018 ESWC program.

The 2014, 2015, 2016 and 2017 editions of the ESWC challenges have been published in books [3,4,8,14] where each challenge, its tasks, evaluation process have been introduced and each system participating to each challenge has been described, detailed and results and comparisons have been shown. The Semantic

---

[3] http://dbpedia.org.

[4] Publicly   accessible   at   https://groups.google.com/forum/#!forum/semantic-sentiment-analysis.

[5] http://www.maurodragoni.com/research/opinionmining/events/.

[6] http://2018.eswc-conferences.org/.

Sentiment Analysis challenge has been included in the four volumes above [9–12]. The 2014 edition of the challenge was also the first edition in parallel with a workshop at ESWC of the same domain that hosted around 20 participants [6]. The 2016 and 2017 editions of the challenge repeated the success of the dual events of the 2014 edition and run in parallel with the Semantic Sentiment Analysis workshop whose proceedings are in the process of publication.

Besides the Semantic Sentiment Analysis challenge described in this chapter and its previous editions, there are a few number of relevant events and challenges that is worth to mention.

SemEval (Semantic Evaluation)[7] consists of a series of evaluations workshops of computational semantic analysis systems. It is now in its eleventh edition[8] and it has been collocated with the 55th annual meeting of the Association for Computational Linguistics (ACL)[9]. Since 2007 the workshop has covered the sentiment analysis topic. During the last edition, SemEval2017 included five tasks for the sentiment analysis track:

- Sentiment Analysis in Twitter. It was subdivided in five subtasks related to message polarity classification, topic-based message polarity classification and tweet quantification. The used languages were English and Arabic and the challenge organizers encouraged to use profile information provided in Twitter such as demographics (e.g. age, location) to analyze its impact on improving sentiment analysis.
- Fine-Grained Sentiment Analysis on Financial Microblogs and News. Divided in two tracks: one related to StockTwits messages consisting of microblog messages focusing on stock market events and assessments from investors and traders, exchanged via the StockTwits microblogging platform and the other related to Twitter messages consisting of tweets about stock market discussion within the Twitter platform. The problem was to predict the sentiment score for each of the companies/stocks mentioned where the sentiment values need to be floating point values in the range of -1 (very negative/bearish) to 1 (very positive/bullish), with 0 designating neutral sentiment.
- #HashtagWars: Learning a Sense of Humor. The goal of this task was to learn to characterize the sense of humor represented in a given show. Given a set of hashtags, the goal was to predict which tweets the show will find funnier within each hashtag. The degree of humor in a given tweet is determined by the labels provided by the show.
- Detection and Interpretation of English Puns. Puns are a class of language constructs in which lexical-semantic ambiguity is a deliberate effect of the communication act. That is, the speaker or writer intends for a certain word or other lexical item to be interpreted as simultaneously carrying two or more separate meanings. The task is divided into three subtasks where puns must be detected, localized and interpreted.

---

[7] https://en.wikipedia.org/wiki/SemEval.

[8] http://alt.qcri.org/semeval2017/.

[9] http://acl2017.org/.

– RumourEval: Determining rumour veracity and support for rumours. This task aimed to identify and handle rumours and reactions to them, in text.

Works such as [2,5,13] represent strong contributions within the domain of semantic sentiment analysis. In those the authors exploited unsupervised techniques to analyse the semantics of a given sentence providing information such as the opinion holder, the topic and the opinion being expressed by the holder to the topic.

Last but not least, authors in [1] provided a feasible research platform for the development of practical solutions for sentiment analysis to be beneficial for our society, business and future research as well.

3   Tasks, Datasets and Evaluation Measures

The fifth edition of the Semantic Sentiment Analysis challenge included three tasks: Polarity Detection and Aspect-Based Sentiment Analysis. One more task was represented by the Most Innovative Approach. Participants had to submit an abstract of no more than 200 words and a 4 pages paper including the details of their systems, why it is innovative, which features or functions it provides, which design choices were made, what lessons were learnt, which tasks it addressed and how the semantics was employed. Industrial tools with non disclosure restrictions were also allowed to participate, and in this case they were asked to:

– explain even at a higher level their approach and engine macro-components, why it is innovative, and how the semantics is involved;
– provide free access (even limited) for research purposes to their engine, especially to make repeatable the challenge results or other experiments possibly included in their paper

As the challenge focused on the introduction, presentation, development and discussion of novel approaches to semantic sentiment analysis, participants had to design a semantic opinion-mining engine that exploited Semantic Web knowledge bases, e.g., ontologies, DBpedia, etc., to perform multi-domain sentiment analysis. Systems not including semantics have been rejected whereas the others had to provide a full description of their system, web access or a link where the system could be downloaded together with a short set of instructions. Moreover, accepted systems had to be either accessible via web or downloadable or anyway a RESTful API had to be provided to run the challenge test-set. If an application was not publicly accessible, password had to be provided for reviewers. A short set of instructions on how to use the application or the RESTFul API had to be provided as well.

Following we will describe each task and, in particular, will detail datasets and evaluation methodologies we have provided for tasks 1 and 2, those targeted by the submitted systems.

```
<?xml version="1.0" encoding="UTF-8" standalone="yes"?>
<Sentences>
    <sentence id="apparel_0">
        <text>
            GOOD LOOKING KICKS IF YOUR KICKIN IT OLD SCHOOL LIKE ME. AND COMFORTABLE.
            AND RELATIVELY CHEAP. I'LL ALWAYS KEEP A PAIR OF STAN SMITH'S
            AROUND FOR WEEKENDS
        </text>
        <polarity>
        positive
        </polarity>
    </sentence>
    <sentence id="apparel_1">
        <text>
            These sunglasses are all right. They were a little crooked, but still cool..
        </text>
        <polarity>
        positive
        </polarity>
    </sentence>
</sentence>
```

Fig. 1. Task 1 output example. Input is the same without the polarity tag.

3.1   Task 1: Polarity Detection

The proposed semantic opinion-mining engines were assessed according to precision, recall and F-measure of the detected polarity values (positive OR negative) for each review of the evaluation dataset. As an example, for the tweet *GOOD LOOKING KICKS IF YOUR KICKIN IT OLD SCHOOL LIKE ME. AND COMFORTABLE. AND RELATIVELY CHEAP. I'LL ALWAYS KEEP A PAIR OF STAN SMITH'S AROUND FOR WEEKENDS*, the correct answer that a sentiment analysis system needed to give was *positive* and therefore it had to write *positive* between the $<polarity>$, $</polarity>$ tags of the output. Figure 1 shows an example of the output schema for task1.

This task was pretty straightforward to evaluate. A precision/recall analysis was implemented to compute the accuracy of the output for this task. A true positive (tp) was defined when a sentence was correctly classified as positive. On the other hand, a false positive (fp) is a positive sentence which was classified as negative. Then, a true negative (tn) is detected when a negative sentence was correctly identified as such. Finally, a false negative (fn) happens when a negative sentence was erroneously classified as positive. With the above definitions, we defined the precision as

$$precision = \frac{tp}{tp + fp}$$

the recall as

$$recall = \frac{tp}{tp + fn}$$

the F1 measure as

$$F1 = \frac{2 \times precision \times recall}{precision + recall}$$

and the accuracy as

$$accuracy = \frac{tp + tn}{tp + fp + fn + tn}$$

As training, development and test sets, we used one million of reviews collected from the Amazon web site and split in 20 different categories: *Amazon Instant Video, Automotive, Baby, Beauty, Books, Clothing Accessories, Electronics, Health, Home Kitchen, Movies TV, Music, Office Products, Patio, Pet Supplies, Shoes, Software, Sports Outdoors, Tools Home Improvement, Toys Games, and Video Games.* The classification of each review (positive or negative) has been done according to the guidelines used for the construction of the Blitzer dataset [7]. Participants evaluated their system by applying a cross-fold validation over the dataset where each fold is clearly delimited. The script to compute Precision, Recall, and F-Measure and the confusion matrix has been provided to participants through the website of the challenge.

```xml
<?xml version="1.0" encoding="UTF-8" standalone="yes"?>
<Review rid="1">
    <sentences>
        <sentence id="348:0">
            <text>Most everything is fine with this machine: speed, capacity, build.</text>
            <Opinions>
                <Opinion aspect="MACHINE" polarity="positive"/>
            </Opinions>
        </sentence>
        <sentence id="348:1">
            <text>The only thing I don't understand is that the resolution of the
            screen isn't high enough for some pages, such as Yahoo!Mail.
            </text>
            <Opinions>
                <Opinion aspect="SCREEN" polarity="negative"/>
            </Opinions>
        </sentence>
        <sentence id="277:2">
            <text>The screen takes some getting use to, because it is smaller
            than the laptop.</text>
            <Opinions>
                <Opinion aspect="SCREEN" polarity="negative"/>
            </Opinions>
        </sentence>
    </sentences>
</Review>
```

Fig. 2. Task 2 output example. Input is the same without the opinion tag and its descendant nodes.

3.2   Task 2: Aspect-Based Sentiment Analysis

Aspect-Based sentiment analysis looks for a binary polarity value associated to aspects extracted from a certain topic. Whereas task 1 ask for an overall polarity value for a, let's say, given review on a hotel, this task asks for a positive or negative value for aspects of the hotel (rooms' quality, cleanness, food, etc.). Submitted systems are evaluated for the aspect extraction and the performed polarity detection through a precision-recall analysis similarly as performed during SemEval 2016 Task 5[10]. Figure 2 shows an example of the output schema for task 2.

The training and test sets were composed by, respectively, 5,058 and 891 sentences coming from three different domains:

- Laptop, (3,048 sentences for training and 728 for testing);
- Restaurant, (2,000 sentences for training);
- Hotel, (163 sentences for testing).

The hotel domain has been chosen to check the efficiency, effectiveness and flexibility of the submitted systems.

3.3   The Most Innovative Approach Task

The system using in the most innovative way common-sense knowledge and semantics would win the most innovative approach task. We would also take into account the usability of the system, the design of the user interface (if applicable), and multi-language capabilities.

4   Submitted Systems

This year we received 9 expressions of interest. The challenge chairs used the Google group mentioned in the introduction section as a forum to explain tasks and requirements needed for the challenge. Details related to the challenge were thus published much ahead of time to let interested researchers know about its status. Thanks to the Google group we have leveraged, there were not any delays during the submission phase. The five submission we received, along their details (title, authors, targeted tasks), are listed in Table 1.

5   Results

A week before the ESWC conference, the evaluation dataset (the one that contained the sentences only) for Task 1 was published. Participants had to run their systems and send to the challenge chairs their results by the next two days. Computing the accuracy was pretty straightforward as accuracy scripts were already prepared and available to download within the website of the challenge. In the following, we will show the results of the participants' systems.

---

[10] http://alt.qcri.org/semeval2016/task5/.

Table 1. The systems participating at the fifth edition of the Semantic Sentiment Analysis challenge and the tasks they targeted.

System	Task 1	Task 3	Most Inn. Approach
Guangyuan Piao and John G. Breslin **Domain-Aware Sentiment Classification with GRUs and CNNs**	X		X
Mattia Atzeni and Diego Reforgiato Recupero **Fine-Tuning of Word Embeddings for Semantic Sentiment Analysis**	X		X
Marco Federici and Mauro Dragoni **The KABSA System at ESWC-2018 Challenge on Semantic Sentiment Analysis**	X	X	X
Mauro Dragoni **The NeuroSent System at ESWC-2018 Challenge on Semantic Sentiment Analysis**	X		X
Mauro Dragoni **The FeatureSent System at ESWC-2018 Challenge on Semantic Sentiment Analysis**	X	X	X
Andi Rexha, Mark Kröll, Mauro Dragoni, and Roman Kern **The CLAUSY System at ESWC-2018 Challenge on Semantic Sentiment Analysis**	X	X	X
Mauro Dragoni and Giulio Petrucci **The IRMUDOSA System at ESWC-2018 Challenge on Semantic Sentiment Analysis**	X		X
Vijayasaradhi Indurthi and Subba Reddy Oota **Evaluating Quality of Word Embeddings with Sentiment Polarity Identification Task**	X		X
Debanjan Chaudhuri and Jens Lehmann **CNN for Product Review Classification with Pre-Trained Word Embeddings**	X		X

5.1   Task 1

In Table 2 we show the precision-recall analysis of the nine systems competing for Task 1. The system of *Guangyuan Piao and John Breslin* had the best f-measure and, therefore, was awarded with a Springer voucher of the value of 100 euros, as the winner of the task. To note that two systems have been disqualified because the output format was not compliant with that provided within the challenge task instructions. That is why they are not included in the table.

In Table 3 we show the precision-recall analysis of the three systems competing for Task 2. The system of *Marco Federici* had the best f-measure and, therefore, was awarded with a Springer voucher of the value of 100 euros, as the winner of the task.

Table 2. Results obtained by the participant to Task #1.

System	F-Measure
Guangyuan et al.	0.9643
Atzori et al.	0.9561
Federici et al.	0.9356
Dragoni et al.	0.9228
Dragoni et al.	0.8823
Rexha et al.	0.8743
Indurthi et al.	0.8203
Petrucci et al.	0.7153
Chaudhuri et al.	0.5243

5.2   Task 2

Table 3. Results obtained by the participant to Task #3.

System	F-Measure
Federici et al.	0.5682
Dragoni et al.	0.5244
Rexha et al.	0.4598

5.3   The Most Innovative Approach Task

This year we did not identified particular disruptive system among the participants. Thus, the Most Innovative Approach award has not been assigned.

6   Conclusions

The Semantic Sentiment Analysis challenge at ESWC2018 followed the success of the first four editions and attracted people within the research and industry world from the semantic web community and traditional Sentiment Analysis and natural language processing techniques. In general, researchers coming from the Sentiment Analysis world become curious and familiar with Semantic Web resources and systems and embed them in their existing methods in order to provide higher accuracy.

Our challenge was coupled with a related workshop where participants were suggested to submit a research work explaining the theory behind their method and how the semantics they employed was effectively used. The workshop was a full day event and attracted around 25 people from research and industry.

This year the organizers of the challenge came from an Italian research institute and a Singaporean university. The two of them are very actively engaged with sentiment analysis research and technology and published and developed several resources, software and papers within that domain. The rationale behind that was to attract researchers and industries from all around the world and have them competing on common tasks of sentiment analysis exploiting Semantic Web technologies. Although only two tasks were targeted, results we obtained of the winning systems were impressive. During the related workshop there was a constructive discussion related to the participants to the challenge and several suggestions were given in order to further improve the precision of those systems, which have also been strongly suggested to participate to other challenge with higher number of participants (e.g. SemEval each year proposes the polarity detection task). Although the number of participants in our challenge were not many, and this is mostly due to the constraints of the Semantic Web technologies that the submitted system had to employ, we aimed at giving advise and suggestions to the few number of participants so that they may compete in other known challenges (e.g. SemEval) with the advantage of exploiting Semantic Web technologies against the systems that still use classical statistical approaches.

We will propose again the dual event challenge-workshop to further provide suggestions and tips to researchers that would like to improve the accuracy of their sentiment analysis methods exploiting Semantic Web technologies and best practices.

Last but not least, we will keep exploiting the Google group we have set up a couple of years ago for dissemination and promotion activities which currently counts 185 members from all around the world.

Acknowledgments. Challenge Organizers want to thank Springer for supporting the provided awards also for this year edition. Moreover, the research leading to these results has received funding from the European Union Horizon 2020 the Framework Programme for Research and Innovation (2014–2020) under grant agreement 643808 Project MARIO Managing active and healthy aging with use of caring service robots.

References

1. Cambria, E., Das, D., Bandyopadhyay, S., Feraco, A.: A Practical Guide to Sentiment Analysis, 1st edn. Springer, Cham (2017). https://doi.org/10.1007/978-3-319-55394-8
2. Consoli, S., Gangemi, A., Nuzzolese, A.G., Recupero, D.R., Spampinato, D.: Extraction of topics-events semantic relationships for opinion propagation in sentiment analysis. In: Proceedings of Extended Semantic Web Conference (ESWC), Crete, GR (2014)
3. Dragoni, M., Solanki, M., Blomqvist, E. (eds.): SemWebEval 2017. CCIS, vol. 769. Springer, Cham (2017). https://doi.org/10.1007/978-3-319-69146-6

4. Gandon, F., Cabrio, E., Stankovic, M., Zimmermann, A. (eds.): SemWebEval 2015. CCIS, vol. 548. Springer, Cham (2015). https://doi.org/10.1007/978-3-319-25518-7

5. Gangemi, A., Presutti, V., Recupero, D.R.: Frame-based detection of opinion holders and topics: a model and a tool. IEEE Comput. Intell. Magaz. **9**(1), 20–30 (2014)

6. Gangemi, A., et al.: Joint proceedings of the 1th workshop on semantic sentiment analysis (ssa2014), and the workshop on social media and linked data for emergency response (smile 2014), co-located with 11th european semantic web conference (eswc 2014), crete, greece, 25 May 2014. http://ceur-ws.org/Vol-1329/

7. Blitzer J., Dredze M., Pereira F.: Biographies, bollywood, boom-boxes and blenders: domain adaptation for sentiment classification. In: Association of Computational Linguistics (ACL) (2007)

8. Presutti, V., et al. (eds.): SemWebEval 2014. CCIS, vol. 475. Springer, Cham (2014). https://doi.org/10.1007/978-3-319-12024-9

9. Recupero, D.R., Cambria, E.: ESWC2014 challenge: concept-level sentiment analysis. In: SemWebEval@ESWC 2014, pp. 3–20, May 2014. http://challenges.2014.eswc-conferences.org/index.php/SemSA

10. Recupero, D.R., Cambria, E., Di Rosa, E.: Semantic sentiment analysis challenge at ESWC2017. In: Dragoni, M., Solanki, M., Blomqvist, E. (eds.) SemWebEval 2017. CCIS, vol. 769, pp. 109–123. Springer, Cham (2017). https://doi.org/10.1007/978-3-319-69146-6_10

11. Recupero, D.R., Dragoni, M., Presutti, V.: ESWC 15 challenge on concept-level sentiment analysis. In: Gandon, F., Cabrio, E., Stankovic, M., Zimmermann, A. (eds.) SemWebEval 2015. CCIS, vol. 548, pp. 211–222. Springer, Cham (2015). https://doi.org/10.1007/978-3-319-25518-7_18

12. Dragoni, M., Recupero, D.R.: Challenge on fine-grained sentiment analysis within ESWC2016. In: Sack, H., Dietze, S., Tordai, A., Lange, C. (eds.) SemWebEval 2016. CCIS, vol. 641, pp. 79–94. Springer, Cham (2016). https://doi.org/10.1007/978-3-319-46565-4_6

13. Recupero, D.R., Presutti, V., Consoli, S., Gangemi, A., Nuzzolese, A.: Sentilo: frame-based sentiment analysis. Cogn. Comput. **7**(2), 211–225 (2014)

14. Sack, H., Dietze, S., Tordai, A., Lange, C. (eds.): SemWebEval 2016. CCIS, vol. 641. Springer, Cham (2016). https://doi.org/10.1007/978-3-319-46565-4

Domain-Aware Sentiment Classification with GRUs and CNNs

Guangyuan Piao[1]($\boxtimes$) and John G. Breslin[2]

[1] Insight Centre for Data Analytics, Data Science Institute, National University of
Ireland Galway, Galway, Ireland
guangyuan.piao@insight-centre.org
[2] IDA Business Park, Galway, Ireland
john.breslin@nuigalway.ie

Abstract. In this paper, we describe a deep neural network architecture
for domain-aware sentiment classification task with the purpose of the
sentiment classification of product reviews in different domains and evaluating nine pre-trained embeddings provided by the semantic sentiment
classification challenge at the 15th Extended Semantic Web Conference.
The proposed approach combines the domain and the sequence of word
embeddings of the summary or text of each review for Gated Recurrent
Units (GRUs) to produce the corresponding sequence of embeddings by
being aware of the domain and previous words. Afterwards, it extracts
local features using Convolutional Neural Networks (CNNs) from the
output of the GRU layer. The two sets of local features extracted from
the domain-aware summary and text of a review are concatenated into a
single vector, and are used for classifying the sentiment of a review. Our
approach obtained 0.9643 F1-score on the test set and achieved the 1st
place in the first task of the Semantic Sentiment Analysis Challenge at
the 15th Extended Semantic Web Conference.

1 Introduction

Sentiment analysis plays an important role in many domains such as predicting
the market reaction in the financial domain [4]. Therefore, many approaches have
been proposed for classifying sentiment labels as well as predicting sentiment
scores. With the advance of deep learning [13] approaches such as Convolutional
Neural Networks (CNNs) [14] for processing data in the form of multiple arrays,
or Recurrent Neural Networks (RNNs) such as Long Short-Term Memory neural
networks (LSTMs) [7] for tasks with sequential inputs, many research areas have
made a significant progress. These research areas include computer vision [5,12],
natural language processing (NLP) [9,10], and recommender systems [19].

Recently, many approaches [3,16] leveraging deep neural networks (DNNs)
have been proposed for sentiment analysis as well. For example, an ensemble
approach using several DNNs such as CNNs and LSTMs proposed by [3] won

© Springer Nature Switzerland AG 2018
D. Buscaldi et al. (Eds.): SemWebEval 2018, CCIS 927, pp. 129–139, 2018.
https://doi.org/10.1007/978-3-030-00072-1_11

one of the sentiment prediction tasks on Twitter[1] at the sentiment analysis challenge SemEval2017[2].

In this paper, we also propose a DNN architecture for the embedding evaluation and semantic sentiment analysis challenge held in conjunction with the 15th Extended Semantic Web Conference[3]. Our proposed approach achieved the best performance compared to the solutions from the other participants in the first task of the challenge.

The first task of the challenge has two main goals as follows:

1. Given nine pre-trained word embeddings and two baseline embeddings, the objective is to evaluate these embeddings in the context of sentiment classification for domain-specific reviews.

2. Propose a sentiment classification model and compare the classification performance on a test set with the models proposed by the other participants.

To this end, the challenge provides a training dataset which consists of one million reviews covering 20 domains. Each training instance consists of a summary, review text, the domain of the review, and its sentiment label (i.e., positive or negative).

Each participated system can train their systems based on the training set and predicts the sentiment labels for the reviews in the test set provided by the organizers. Those systems are evaluated by the F1-score on the test set, which can be defined as follows:

$$F1 = \frac{2 \times precision \times recall}{precision + recall} \tag{1}$$

$$precision = \frac{TP}{TP + FP} \tag{2}$$

$$recall = \frac{TP}{TP + FN} \tag{3}$$

where $tp \in TP$ (true positive) is defined when a sentence is correctly classified as positive, and $fp \in FP$ (false positive) denotes when a positive review is classified as negative. A true negative ($tn \in TN$) denotes when a negative sentence is correctly classified as negative, and a false negative ($fn \in FN$) denotes the case of a negative sentence is classified as positive.

2 System Description

In this section, we describe the architecture of our proposed system using domain-aware GRUs and CNNs with pre-trained word embeddings for the sentiment classification of reviews. Figure 1 provides an overview of our proposed model for the semantic sentiment analysis challenge.

---

[1] https://twitter.com.
[2] http://alt.qcri.org/semeval2017/index.php?id=tasks.
[3] https://2018.eswc-conferences.org/.

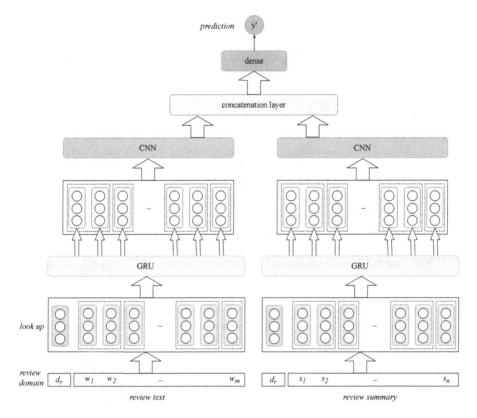

Fig. 1. The proposed system architecture for predicting the sentiment of a given review. A review is represented as $r = \{s, x, d_r\}$ where s, x, and d_r denote the summary, review text, and the domain of r, respectively. First, we represent the text x of a review as a sequence of embeddings: $\{E_{d_r}, E_{w_1}, \ldots, E_{w_m}\}$ where E_{d_r} denotes the domain embedding of r, and $E_{w_1}, \ldots, E_{w_m}$ denote the set of word embeddings for the sequence of words in x. Afterwards, the sequence of embeddings is feed into the GRU layer. The output of this GRU layer is a sequence of embeddings with the memory of the domain as well as previous words, and is used as an input to a CNN layer to extract local features of the text x. Similarly, we extract the local features of the summary of r. Finally, the local features of the text and summary of a review are concatenated, and fed into a dense layer in order to predict the sentiment label of r.

We represent each review r as two parts using the review text and summary. For example, the first part consists of the domain of r (d_r), and the sequence of words $\{w_1, \ldots, w_m\}$ in the text of r. The second part consists of the domain of r (d_r), and the sequence of words $\{s_1, \ldots, s_n\}$ in the summary of r. Each training instance can be represented as $t_i = \{r_i, y_i\}$ where y denotes the sentiment label (i.e., 1 or 0) of r. As we can see from Fig. 1, there are mainly four layers in our proposed model. The first part of review with its domain and text goes through the following layers.

Look up. The look up layer transforms a review r into the domain and the sequence of word embeddings: $\{E_{d_r}, E_{w_1}, \ldots, E_{w_m}\}$ where E_{d_r} denotes the domain embedding of r, and $E_{w_1}, \ldots, E_{w_m}$ denote the set of pre-trained word embeddings for the sequence of words in r's text. The domain embeddings are also parameters of our model and will be learned through the training process.

GRU layer. The sequence of embeddings $\{E_{d_r}, E_{w_1}, \ldots, E_{w_m}\}$ for a review from the look up layer is used as an input to a Gated Recurrent Unit [2] (GRU) layer. This layer aims to transform the domain and word embeddings of a review r into a sequence of embeddings (hidden states in GRUs) for r where each embedding at position k is aware of the domain as well as the previous words. GRUs and LSTMs are RNNs that have been designed to resolve the issue of vanishing gradient problem in the vanilla RNNs. We opted to use GRUs instead of LSTMs as the former one has fewer number of parameters to tune, which is suitable for the limited computing resources with a large number of training instances. We provide more details about GRU in Sect. 2.1.

CNN layer. Recurrent Neural Networks such as GRU is good at capturing the long-term dependencies of words, e.g., capturing sarcasm of a text. In contrast, the CNN layer aims to extract local features which are useful for the sentiment classification. CNNs apply a set of convolutional filters to the input matrix with a filtering matrix $f \in \mathbb{R}^{h \times d}$ where h is the filter size. This filter size denotes the number of the output embeddings from the GRU layer it spans. In our proposed model, we used $\{3, 4, 5\}$ as the filter sizes and 200 filters for the first part of a review with its review text. For the second part of a review with its review summary, we used $\{1, 2, 3\}$ as the filter sizes and 100 filters in total. We use C_t and C_s to denote the local features extracted from the two parts of a review with its text and summary. We provide more details about CNNs in Sect. 2.2.

Concatenation layer. Similar to the first part of a review, the second part of it with its summary also goes through the abovementioned layers. In the concatenation layer, the outputs from both parts are concatenated together to represent the review: $C = [C_s, C_t]$. This vector of the review is used as an input to a dense layer.

Dense layer. Finally, there is a dense layer with 80 hidden nodes (D) to shrink the dimension of C to 80 dimensions. The dense layer is then used for classifying the sentiment of a review.

$$y' = W \cdot D + b \tag{4}$$

where W is the weight parameters for this dense layer and b denotes a bias term.

Loss function. We use the cross-entropy as our loss function, and the objective is to minimize the loss over all training examples.

$$L = \sum_i (y_i \times -log(\sigma(y'_i)) + (1 - y_i) \times -log(1 - \sigma(y'_i))) \tag{5}$$

where σ is the *sigmoid* function, $\sigma(y'_i)$ is the predicted sentiment score, and y_i is the ground truth sentiment label of i-th training instance.

2.1   Gated Recurrent Units

Gated Recurrent Units along with LSTMs are part of the RNN family, which is designed to deal with sequential data. For each element of the sequence (e.g., each word of a review), a vanilla RNN uses the current element embedding and its previous hidden state to output the next hidden state:

$$h_t = f(W_h \cdot x_t + U_h \cdot h_{t-1} + b_h) \tag{6}$$

where $h_t \in \mathbb{R}^m$ denotes the hidden state at t-th element (or at time step t), x_t is the current element embedding, $W_h \in \mathbb{R}^{m \times d}$ and $U_h \in \mathbb{R}^{m \times m}$ are weight matrices, b_h denotes a bias term, and $f(\cdot)$ is a non-linear activation function such as $tanh$. One main problem of this vanilla RNN is that it suffers from the vanishing gradient problem.

This problem has been solved in LSTMs with a memory cell instead of a single activation function. GRUs (see Fig. 2) can be seen as a variation on the LSTMs but with a smaller number of parameters, and both GRUs and LSTMs produce comparative results in many problems. To resolve the vanishing gradient problem of a vanilla RNN, GRUs use *update* and *reset* gates, which control the information to be passed to the output. The formula for the output gate is as follows:

$$z_t = \sigma(W_z \cdot x_t + U_z \cdot h_{t-1}) \tag{7}$$

where W_z and U_z are weight matrices and σ denotes the *sigmoid* function. This gate determines how much information from the past can be passed along to the

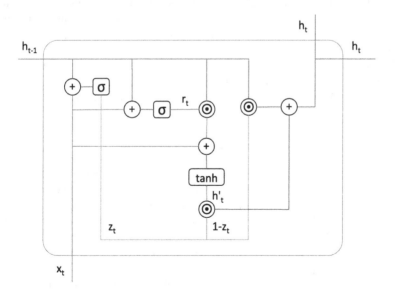

Fig. 2. The architecture of a Gradient Recurrent Unit based on the figure from https://towardsdatascience.com/understanding-gru-networks-2ef37df6c9be.

future. In contrast, the reset gate decides how much information from the past should be forgotten using the formula as below.

$$r_t = \sigma(W_r \cdot x_t + U_r \cdot h_{t-1}) \tag{8}$$

where W_r and U_r are weight matrices of the reset gate. The output of the reset gate is used to calculate the current memory content as follows:

$$h'_t = tanh(W_h \cdot x_t + r_t \odot U_h \cdot h_{t-1}) \tag{9}$$

where W_h and U_h are weight matrices. The element-wise product of r_t (which ranges from 0 to 1) and $U_h \cdot h_{t-1}$ determines how much information to remove from the previous time steps.

Finally, the current hidden state is determined by using the current memory content h'_t and the previous hidden state h_{t-1} with the update gate.

$$h_t = (1 - z_t) \odot h'_t + z_t \odot h_{t-1} \tag{10}$$

Based on these formulas, we can observe that the vanilla RNN can be see as a special case of a GRU when $z_t = 0$ and $r_t = 1$.

2.2 Convolutional Neural Networks

Here we give an overview of CNNs. A smaller version of the used model is presented in Fig. 3. The input of CNNs in our model is the output of the GRU layer, which consists of a sequence of hidden states. We use GRU matrix $G \in \mathbb{R}^{m \times d}$ to denote this sequence of hidden states. In contrast to using the sequence of word embeddings as an input, using those hidden states obtained through the GRU layer embodies the information from previous words as well as the domain of a review.

We then apply a set of convolutional filters to the GRU matrix with a filtering matrix $f \in \mathbb{R}^{h \times d}$ where h is the filter size which denotes the number of hidden states it spans. The convolutional operation is defined as follows:

$$c_i = f\left(\sum_{j,k} w_{j,k}(G_{[i:i+h-1]})_{j,k} + b_c\right) \tag{11}$$

where $b_c \in \mathbb{R}$ is a bias term, and $f(\cdot)$ is a non-linear function (we use the $ReLu$ function in our model). The output $c \in \mathbb{R}^{m-h+1}$ is a concatenated vector which consists of c_i over all windows of hidden states in G. We can define the number of filtering matrices which can be applied to learn different features. As mentioned earlier, we used $\{3, 4, 5\}$ as the filter sizes and 200 filters for the first part of a review with its review text. For the second part of a review with its review summary, we used $\{1, 2, 3\}$ as the filter sizes and 100 filters in total.

In order to extract the most important feature for each convolution, we use the $max - pooling$ operation $c_{max} = max(c)$. Therefore, this operation extracts the most important feature with regardless of the location of this feature in the

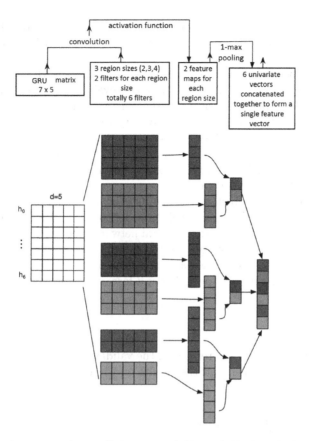

Fig. 3. The architecture of a smaller version of the used CNNs. Figure is taken from [20] with some modifications.

sequence of hidden states. Afterwards, the c_{max} of each filter is concatenated into a single vector $\mathbf{c}_{max} \in \mathbb{R}^p$ where p is the total number of filters. For example, the total number of filters for the first part of a review is $3 \times 200 = 600$, and that for the second part of the review is $3 \times 100 = 300$ in our model. These two vectors are concatenated together in the concatenation layer as we can see from the Fig. 1 which is followed by a dense layer.

2.3 Training

This section provides some details about training our proposed model. The parameters to be learned through the training process include the domain embeddings, the parameters associated with GRUs and CNNs as well as the nodes in the dense layer.

Validation set. In order to train our proposed model, we used 10,000 out of the 1 million provided reviews which are evenly distributed over 20 domains as the validation set. Therefore, the validation set consists of 500 training instances for each domain where 250 of them are positive ones and the rest are negative ones.

Training. We train our proposed model nine times with each of the nine word embeddings provided. As we also aim to evaluate these pre-trained word embeddings, we do not further optimize the weights of these embeddings, i.e., the weights of word embeddings are fixed. To learn the parameters of our proposed approach for minimizing the loss, we use a mini-batch gradient decent with 128 as the batch size and use the Adam update rule [11] to train the model on the training set, and stop the training if there is no improvement on the validation set within the subsequent 10 epochs.

Regularization. We further use the dropout [18] for regularization with a dropout rate $p = 0.5$. Dropout is one of the widely used regularization approaches for preventing overfitting in training neural network models. In short, individual nodes are "disabled" with probability $1 - p$, and then the outputs with the set of disabled nodes of a hidden layer are used as an input to the next layer. In this way, it prevents units from co-adapting and forces them to learn useful features individually.

3 Results

In this section, we discuss the results on the validation set using our model with different word embeddings, and the results on the test set compared to all the participated systems.

3.1 Results on the Validation Set

Table 1 shows the best classification accuracy ($accuracy = \frac{TP+TN}{TP+FP+TN+FN}$) on the validation set using the nine provided embeddings and the two baseline ones using our proposed architecture. The provided embeddings are denoted as emb_{dimensionality of word embeddings}_{epochs for training}.

Instead of using each model with different word embeddings, we further investigated an ensemble approach which uses a simple voting scheme for 11 variants of our proposed model with different word embeddings or training epochs. The final sentiment classification is determined by the votes from the 11 classification results where the sentiment label with a higher number of votes is selected. Based on the results, we have the following observations:

- The classification performance with the nine provided embeddings outperforms the one with the two baseline embeddings.
- The accuracy of sentiment classification increases with a higher dimensionality or number of epochs of the provided word embeddings.
- Overall, emb_512_50, which uses 512 as the dimensionality of word embeddings and was trained for 50 epochs, provides the best performance among all embeddings.
- The simple ensemble approach improves the classification accuracy further compared to using a single word embeddings.

Table 1. The accuracy of sentiment classification on the validation set with the nine provided embeddings.

Embedding	Accuracy
Glove	0.9525
amazon_we	0.9122
emb_128_15	0.9578
emb_128_30	0.9560
emb_128_50	0.9544
emb_256_15	0.9576
emb_256_30	0.9576
emb_256_50	0.9578
emb_512_15	0.9568
emb_512_30	0.9598
emb_512_50	0.9605
Ensemble approach	**0.9657**

Table 2. The comparison of our approach and the other participated ones for the sentiment classification on the test set.

Year	Ranking	F1-score
2018	**#1 (our approach)**	**0.9643**
	#2	0.9561
	#3	0.9356
	#4	0.9228
	#5	0.8823
	#6	0.8743
	#7	0.7153
	#8	0.5243
2017	#1 [1]	0.8675
	#2 [6]	0.8424
	#3 [8]	0.8378
	#4 [15]	0.8112

3.2 Results on the Test Set

The final test set released by the challenge organizers consists of 10,000 examples, which are evenly distributed over 20 domains. Table 2 shows the sentiment classification results on the test set from all the participated systems in this year and those in the last year [17], which are released by the challenge organizers.

As we can see from the table, our proposed approach provides the best performance among all participants in terms of F1-score. In addition, we observe that the performance of the participated systems for this year is significantly improved compared to that for the previous year.

4   Conclusions

In this paper, we presented our system architecture for the Semantic Sentiment Analysis challenge at the 15th Extended Semantic Web Conference. Our goal was to use and compare pre-trained word embeddings with proposed architecture, which combines the advantages of both GRUs and CNNs for classifying the sentiment labels of domain-specific reviews. For future work, the current model can be improved by using bi-directional GRUs to learn the hidden states of GRUs based on both previous and post information. In addition, the ensemble approach with a hard voting scheme can be replaced by another ensemble approach such as a soft voting scheme.

Acknowledgments. This publication has emanated from research conducted with the financial support of Science Foundation Ireland (SFI) under Grant Number SFI/12/RC/2289 (Insight Centre for Data Analytics).

References

1. Atzeni, Mattia, Dridi, Amna, Reforgiato Recupero, Diego: Fine-grained sentiment analysis on financial microblogs and news headlines. In: Dragoni, Mauro, Solanki, Monika, Blomqvist, Eva (eds.) SemWebEval 2017. CCIS, vol. 769, pp. 124–128. Springer, Cham (2017). https://doi.org/10.1007/978-3-319-69146-6_11
2. Cho, K., et al.: Learning phrase representations using RNN encoder-decoder for statistical machine translation. arXiv preprint arXiv:1406.1078 (2014)
3. Cliche, M.: BB_twtr at SemEval-2017 Task 4: twitter Sentiment Analysis with CNNs and LSTMs. CoRR abs/1704.0 (2017), arXiv:1704.06125
4. Cortis, K., et al.: Semeval-2017 task 5: fine-grained sentiment analysis on financial microblogs and news. In: Proceedings of the 11th International Workshop on Semantic Evaluation (SemEval-2017), pp. 519–535 (2017)
5. Farabet, C., Couprie, C., Najman, L., LeCun, Y.: Learning hierarchical features for scene labeling. IEEE Trans. Pattern Anal. Mach. Intell. **35**(8), 1915–1929 (2013)
6. Federici, Marco, Dragoni, Mauro: A knowledge-based approach for aspect-based opinion mining. In: Sack, Harald, Dietze, Stefan, Tordai, Anna, Lange, Christoph (eds.) SemWebEval 2016. CCIS, vol. 641, pp. 141–152. Springer, Cham (2016). https://doi.org/10.1007/978-3-319-46565-4_11
7. Hochreiter, S., Schmidhuber, J.: Long short-term memory. Neural Comput. **9**(8), 1735–1780 (1997). https://doi.org/10.1162/neco.1997.9.8.1735
8. Iguider, Walid, Reforgiato Recupero, Diego: Language independent sentiment analysis of the shukran social network using apache spark. In: Dragoni, Mauro, Solanki, Monika, Blomqvist, Eva (eds.) SemWebEval 2017. CCIS, vol. 769, pp. 129–132. Springer, Cham (2017). https://doi.org/10.1007/978-3-319-69146-6_12
9. Kalchbrenner, N., Grefenstette, E., Blunsom, P.: A convolutional neural network for modelling sentences. In: The 52nd Annual Meeting of the Association for Computational Linguistics (2014)

10. Kim, Y.: Convolutional neural networks for sentence classification. In: Conference on Empirical Methods on Natural Language Processing (2014)
11. Kingma, D., Ba, J.: Adam: a method for stochastic optimization. arXiv preprint arXiv:1412.6980 (2014)
12. Krizhevsky, A., Sutskever, I., Hinton, G.E.: Imagenet classification with deep convolutional neural networks. In: Advances in Neural Information Processing Systems, pp. 1097–1105 (2012)
13. LeCun, Y., Bengio, Y., Hinton, G.: Deep learning. Nature **521**(7553), 436–444 (2015)
14. LeCun, Y., et al.: Handwritten digit recognition with a back-propagation network. In: Advances in Neural Information Processing Systems, pp. 396–404 (1990)
15. Petrucci, Giulio, Dragoni, Mauro: The IRMUDOSA system at ESWC-2017 challenge on semantic sentiment analysis. In: Dragoni, Mauro, Solanki, Monika, Blomqvist, Eva (eds.) SemWebEval 2017. CCIS, vol. 769, pp. 148–165. Springer, Cham (2017). https://doi.org/10.1007/978-3-319-69146-6_14
16. Piao, G., Breslin, J.G.: Financial aspect and sentiment predictions with deep neural networks: an ensemble approach. In: Financial Opinion Mining and Question Answering Workshop at The Web Conference (WWW). ACM (2018)
17. Reforgiato Recupero, Diego, Cambria, Erik, Di Rosa, Emanuele: Semantic sentiment analysis challenge at ESWC2017. In: Dragoni, Mauro, Solanki, Monika, Blomqvist, Eva (eds.) SemWebEval 2017. CCIS, vol. 769, pp. 109–123. Springer, Cham (2017). https://doi.org/10.1007/978-3-319-69146-6_10
18. Srivastava, N., Hinton, G.E., Krizhevsky, A., Sutskever, I., Salakhutdinov, R.: Dropout: a simple way to prevent neural networks from overfitting. J. Mach. Learn. Res. **15**(1), 1929–1958 (2014)
19. Zhang, S., Yao, L., Sun, A.: Deep learning based recommender system: a survey and new perspectives. CoRR abs/1707.0 (2017), arXiv:1707.07435
20. Zhang, Y., Wallace, B.C.: A sensitivity analysis of (and practitioners' guide to) convolutional neural networks for sentence classification. CoRR abs/1510.0 (2015), arXiv:1510.03820

Fine-Tuning of Word Embeddings
for Semantic Sentiment Analysis

Mattia Atzeni[(⊠)] and Diego Reforgiato Recupero

Department of Mathematics and Computer Science, University of Cagliari, Via
Ospedale 72, 09124 Cagliari, Italy
m.atzeni38@studenti.unica.it, diego.reforgiato@unica.it

Abstract. In this paper, we present a state-of-the-art deep-learning
approach for sentiment polarity classification. Our approach is based on
a 2-layer bidirectional Long Short-Term Memory network, equipped with
a neural attention mechanism to detect the most informative words in
a natural language text. We test different pre-trained word embeddings,
initially keeping these features frozen during the first epochs of the training process. Next, we allow the neural network to perform a fine-tuning of
the word embeddings for the sentiment polarity classification task. This
allows projecting the pre-trained embeddings in a new space which takes
into account information about the polarity of each word, thereby being
more suitable for semantic sentiment analysis. Experimental results are
promising and show that the fine-tuning of the embeddings with a neural
attention mechanism allows boosting the performance of the classifier.

Keywords: Word embeddings · LSTM · Deep Learning · Neural
attention · Sentiment Analysis · RNN · Natural Language Processing

1 Introduction

Over the last few years, Deep Learning techniques and algorithms have been
applied in a wide range of fields, leading to a remarkable improvement of the
state of the art in different scenarios. Such approaches have also been proved to
be suitable for well-known Natural Language Processing (NLP) tasks, such as
Named Entity Recognition (NER), part of speech (POS) tagging and Semantic
Role Labeling (SRL) [5]. Moreover, DL techniques have been applied for Gener-
ative Conversational Agents, Sentiment Analysis and Human-Robot interaction
[2].

In this context, word embeddings [12] are certainly among the most significant improvements related to Deep Learning in NLP. Hence, the ESWC 2018
Challenge on Semantic Sentiment Analysis[1] aims at advancing research on opinion mining, emotion detection and sentiment polarity classification, by providing
high-quality resources and pre-trained word embeddings. Following this research

---

[1] http://www.maurodragoni.com/research/opinionmining/events/challenge-2018/.

© Springer Nature Switzerland AG 2018
D. Buscaldi et al. (Eds.): SemWebEval 2018, CCIS 927, pp. 140–150, 2018.
https://doi.org/10.1007/978-3-030-00072-1_12

line, in this paper we describe a state-of-the-art system for sentiment polarity classification [8,16,18]. Our approach is totally based on word embeddings and Deep Learning technologies and it has competed at the ESWC 2018 Challenge on Semantic Sentiment Analysis with promising results.

However, using pre-trained word embeddings still poses some limitations on the capabilities of the approach, since the system should be able to implicitly learn the most suitable features. On the one hand, one common solution to this problem provides for the generation of the word embeddings on a domain-related corpus. On the other hand, when the corpus is not sufficiently large, this process results in a higher number of out-of-dictionary words. Hence, a trade-off between quality and coverage often needs to be found.

To address this issue, we propose a neural network architecture capable of performing a fine-tuning of pre-trained embeddings, considering information about the class labels. This allows improving the quality of the model, while keeping the same coverage given by the pre-trained word embeddings. Our approach relies on a 2-layer Bidirectional Long Short-Term Memory (LSTM) network which has been equipped with a neural attention mechanism. Experimental results are promising and show that the fine-tuning of the embeddings allows boosting the performance of the classifier.

The remainder of this paper is organized as follows. Section 2 provides an extensive overview the problem that is tackled in the paper and gives a description of the dataset supplied by the organizers of the challenge. Section 3 discusses the architecture of the proposed system and its main advantages with respect to other state-of-the-art approaches, such as CNNs and ensemble models. Section 4 digs into details about the pre-processing of the data, the training process and the choice of the hyper-parameters. Finally, Sect. 4 deals with the evaluation of the system, while Sect. 6 concludes the paper.

2   Task Overview

Our system competed in the first task of the challenge, which aimed at classifying a natural language text with a binary label, denoting whether it conveyed a positive or negative opinion. The training set, provided by the organizers of the challenge, consists of 1 million reviews extracted from Amazon. Each record has the following attributes:

- **id:** a unique identifier of the review;
- **domain:** each review falls into one of 20 different domains (for more details see the URL of the challenge);
- **summary:** a natural language summary of the content of the review;
- **text:** the actual text of the review;
- **polarity:** a class label denoting whether the review conveys a "positive" or "negative" opinion.

Overall, the training set is balanced, with 500000 positive and as many negative reviews. The structure of the test set is the same, without the polarity label.

The organizers of the challenge also provided several sets of pre-trained word embeddings, that each participating system was required to use[2]. The supplied embeddings have different sizes and have been generated using different training epochs. The size of the embeddings can either be 128, 256 or 512, while the training process has been carried out for 15, 30 or 50 epochs. Hence, we have a total of 9 sets of pre-trained word embeddings that can be used as a starting point for higher-level polarity detection purposes. Furthermore, the organizers of the challenge also provided some baseline embeddings, including the well-known 300-dimensional embeddings generated by applying the Word2Vec algorithm on Google News [12].

Clearly, this enabled validating not only the quality of participating systems, but also which combination of size and training epochs worked better for the pre-trained embeddings. However, in our system, the final classification does not strictly rely on the provided embeddings. Indeed, such vectors are only used during the training process, because our system is capable of changing the embeddings, taking into account information about the polarity of the text provided as training. This allows producing a new set of word embeddings that is more suitable for the polarity detection task and that has been used to classify the reviews in the test set.

3   System Architecture

Many top-ranked approaches in Sentiment Analysis competitions have employed Convolutional Neural Networks (CNNs) [11] for binary polarity classification [13]. Such systems perform well, because a CNN implicitly learns the most suitable features. However, in NLP tasks, a CNN is not capable of effectively leveraging essential information about the order of the words and, therefore, recent research is also focusing on different approaches.

A common alternative to CNNs is represented by Recurrent Neural Networks (RNNs), that are capable of handling a sequential input. More precisely, let $\mathbf{x}_t \in \mathbb{R}^n$ and $\mathbf{h}_t \in \mathbb{R}^m$ be respectively the input of the network and its hidden state at timestep t. Also, let $\mathbb{F}^{a,b}$ be the set of all affine transformations from $\mathbb{R}^a$ to $\mathbb{R}^b$. In the following, we will use the notation $F^{a,b} : \mathbb{R}^a \to \mathbb{R}^b$ to refer to such a kind of transformation, i.e. functions of the form $\mathbf{x} \mapsto A\mathbf{x} + \mathbf{b}$ for some A and $\mathbf{b}$. Then, a traditional RNN can be described using deterministic transitions from current to previous hidden states: $(\mathbf{x}_t, \mathbf{h}_{t-1}) \mapsto \mathbf{h}_t$, and, more precisely, this mapping is given by

$$\mathbf{h}_t = f(F_x^{n,m}(\mathbf{x}_t) + F_h^{m,m}(\mathbf{h}_t)),$$

where f is usually chosen to be the sigmoid function or the hyperbolic tangent. Thus, the hidden state of the network at timestep t depends on the previous hidden states, thereby allowing the RNN to keep a notion of the order of the words in a natural language text.

---

[2] http://www.maurodragoni.com/research/opinionmining/dranziera/embeddings-evaluation.php.

One of the main problems related to RNNs is that learning dependencies over long time windows is difficult, as the gradient may explode or vanish. Several solutions to this problem have been proposed, including the well-known Long Short-Term Memory [9] (LSTM) units, which have been recently employed also for Sentiment Analysis tasks [3]. One of the main drawbacks of such models is that they are less effective at performing feature extraction with respect to CNNs. To overcome this issue, some systems are actually combining both RNNs and CNNs. For instance, in [21] a single architecture employing both convolutional and gated recurrent neural networks is used to perform sentiment analysis on Twitter. Also, the top-scoring system of SemEval 2017 Task 4 [19] used an ensemble of 10 CNNs and 10 LSTMs tuned with different hyper-parameters and embeddings [4].

However, an ensemble strategy is well suited for competitions, but has some drawbacks in real-world applications, as the training and classification times increase with the number of models. To tackle this problem, we propose a single model which relies solely on LSTM units, while keeping the ability of a CNN of effectively learning the most suitable features. More in details, we experiment a fine-tuning of the word embeddings for the polarity classification task. Our model is capable of modifying the pre-trained word embeddings during the training process, in order to apply the new set of embeddings for classification purposes.

The architecture is based on a 2-layer Bidirectional LSTM, equipped with a neural attention mechanism, as shown in Fig. 1.

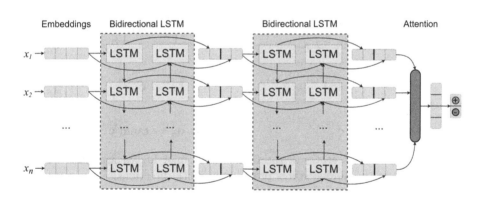

Fig. 1. High-level view of the architecture of the RNN.

Let T be the set of all the tokens in the dataset and let $id : T \rightarrow [1, |T|]$ be a bijective function associating a unique identifier to each token $a \in T$. The input of the system is a natural language message, that is encoded as a sequence $\mathbf{x} = (x_1, \ldots, x_n)$, where n is the maximum length of the text and $x_t = id(a_t)$ is the unique identifier of the t-th token in the message, a_t, for each $t = 1, \ldots, n$.

The network makes use of an embedding layer that implements a mapping $[1, |T|] \rightarrow \mathbb{R}^m$, where m is the size of the word embeddings. This layer is used to project each identifier x_t in the corresponding vector $\mathbf{v}_t \in \mathbb{R}^m$, $t = 1, \ldots, n$. The weights of the layer are initialized with the pre-trained word embeddings provided by the organizers of the challenge.

The embeddings resulting from this process are fed into two Bidirectional LSTM layers, to obtain an annotation $\mathbf{h}_t$, for each vector $\mathbf{v}_t$, $t = 1, \ldots, n$. A bidirectional LSTM consists of a forward LSTM, that processes the message from the first to the last token, and a backward LSTM that reads the text in the reverse order. More formally, let l be the dimensionality of the output space of the LSTM, σ the sigmoid function, tanh the hyperbolic tangent and $\circ$ the Hadamard product. The hidden states of the first forward and backward LSTMs, denoted as $\overrightarrow{\mathbf{h}}_t^{(1)}$ and $\overleftarrow{\mathbf{h}}_t^{(1)}$ respectively, are computed as:

$$\overrightarrow{\mathbf{i}}_t^{(1)} = \sigma\left(\overrightarrow{F}_i^{l+m,l}\begin{pmatrix}\mathbf{v}_t \\ \mathbf{h}_{t-1}\end{pmatrix}\right), \qquad \overleftarrow{\mathbf{i}}_t^{(1)} = \sigma\left(\overleftarrow{F}_i^{l+m,l}\begin{pmatrix}\mathbf{v}_t \\ \mathbf{h}_{t+1}\end{pmatrix}\right),$$

$$\overrightarrow{\mathbf{g}}_t^{(1)} = \tanh\left(\overrightarrow{F}_g^{l+m,l}\begin{pmatrix}\mathbf{v}_t \\ \mathbf{h}_{t-1}\end{pmatrix}\right), \qquad \overleftarrow{\mathbf{g}}_t^{(1)} = \tanh\left(\overleftarrow{F}_g^{l+m,l}\begin{pmatrix}\mathbf{v}_t \\ \mathbf{h}_{t+1}\end{pmatrix}\right),$$

$$\overrightarrow{\mathbf{f}}_t^{(1)} = \sigma\left(\overrightarrow{F}_f^{l+m,l}\begin{pmatrix}\mathbf{v}_t \\ \mathbf{h}_{t-1}\end{pmatrix}\right), \qquad \overleftarrow{\mathbf{f}}_t^{(1)} = \sigma\left(\overleftarrow{F}_f^{l+m,l}\begin{pmatrix}\mathbf{v}_t \\ \mathbf{h}_{t+1}\end{pmatrix}\right),$$

$$\overrightarrow{\mathbf{o}}_t^{(1)} = \sigma\left(\overrightarrow{F}_o^{l+m,l}\begin{pmatrix}\mathbf{v}_t \\ \mathbf{h}_{t-1}\end{pmatrix}\right), \qquad \overleftarrow{\mathbf{o}}_t^{(1)} = \sigma\left(\overleftarrow{F}_o^{l+m,l}\begin{pmatrix}\mathbf{v}_t \\ \mathbf{h}_{t+1}\end{pmatrix}\right),$$

$$\overrightarrow{\mathbf{c}}_t^{(1)} = \overrightarrow{\mathbf{f}}_t^{(1)} \circ \overrightarrow{\mathbf{c}}_{t-1}^{(1)} + \overrightarrow{\mathbf{i}}_t^{(1)} \circ \overrightarrow{\mathbf{g}}_t^{(1)}, \qquad \overleftarrow{\mathbf{c}}_t^{(1)} = \overleftarrow{\mathbf{f}}_t^{(1)} \circ \overleftarrow{\mathbf{c}}_{t+1}^{(1)} + \overleftarrow{\mathbf{i}}_t^{(1)} \circ \overleftarrow{\mathbf{g}}_t^{(1)},$$

$$\overrightarrow{\mathbf{h}}_t^{(1)} = \overrightarrow{\mathbf{o}}_t^{(1)} \circ \tanh(\overrightarrow{\mathbf{c}}_t^{(1)}), \qquad \overleftarrow{\mathbf{h}}_t^{(1)} = \overleftarrow{\mathbf{o}}_t^{(1)} \circ \tanh(\overleftarrow{\mathbf{c}}_t^{(1)}).$$

Then, the final hidden states of the first bidirectional LSTM are given by the concatenation of the hidden states provided by both directions:

$$\mathbf{h}_t^{(1)} = \begin{pmatrix}\overrightarrow{\mathbf{h}}_t^{(1)} \\ \overleftarrow{\mathbf{h}}_t^{(1)}\end{pmatrix}, \qquad t = 1, \ldots, n.$$

Such annotations are fed into a second bidirectional LSTM, that similarly computes

$$\mathbf{h}_t = \mathbf{h}_t^{(2)} = \begin{pmatrix}\overrightarrow{\mathbf{h}}_t^{(2)} \\ \overleftarrow{\mathbf{h}}_t^{(2)}\end{pmatrix}, \qquad t = 1, \ldots, n,$$

where $\overrightarrow{\mathbf{h}}_t^{(2)} \in \mathbb{R}^l$ and $\overleftarrow{\mathbf{h}}_t^{(2)} \in \mathbb{R}^l$ denote respectively the hidden states of the second-layer forward and backward LSTMs.

Next, we apply a neural attention mechanism [15], to detect the most informative words in the input text and assign a weight w_t to the annotations provided by the second bidirectional LSTM, for each $t = 1, \ldots, n$. This allows computing a final vector $\mathbf{r} \in \mathbb{R}^{2l}$, corresponding to the whole natural language review, as the weighted sum of all the contributions provided by the hidden states of the RNN. More in details, we compute:

$$\mathbf{s} = \tanh\left(F_s^{2ln,n}\begin{pmatrix}\mathbf{h}_1 \\ \vdots \\ \mathbf{h}_n\end{pmatrix}\right), \qquad \mathbf{s} \in \mathbb{R}^n,$$

$$\mathbf{w} = \frac{\exp(\mathbf{s})}{\mathbf{1}^\mathsf{T} \cdot \exp(\mathbf{s})}, \qquad \sum_{t=1}^{n} w_t = 1,$$

$$\mathbf{r} = \sum_{t=1}^{n} w_t \cdot \mathbf{h}_t.$$

Finally, the vector $\mathbf{r}$ is fed into a fully-connected softmax layer that provides the probability distribution over the classes.

4   Training

The model has been trained on the dataset provided by the organizers of the challenge and described in Sect. 2. We represent a review as a natural language message obtained by concatenating the summary and the text fields, and we apply the following pre-processing steps:

- each review is converted to lowercase;
- URLs are normalized and replaced by the url token;
- characters that occur more than 2 times in a row are replaced by 2 repetitions of that letter;
- each review is tokenized into a list of individual words.

Next, each list of tokens is converted into a sequence of integers, as explained in Sect. 3. We set the maximum size of a review to $n = 256$. Reviews that exceed this length are truncated, whereas sequences whose length is less than n are padded with zeros.

Overall, the dataset and the supplied test set contain roughly 400 thousand tokens, whereas each provided set of word embeddings has more than 4 million vectors. Hence, in order to improve loading times and reduce memory requirements, we pre-process the word embeddings by removing vectors that are associated to words that do not occur in the dataset. This reduced set of embeddings is then used to initialize the weights of the embedding layer.

As a regularization technique, we apply Gaussian noise to the embeddings, with a standard deviation of 0.2. This process can be seen as a randomized data augmentation, that enhances the performance of the model while reducing the risk of overfitting. Also, we make extensive use of dropout [20] after the embedding layer and the bidirectional LSTMs. Moreover, dropout is also added at the recurrent connections of the LSTMs.

We set the output dimensionality of the LSTM to $l = 64$ and the network is trained to instantiate the affine transformations defined in Sect. 3, in order to minimize the categorical cross-entropy loss function. The training process relies on the back-propagation algorithm with stochastic gradient descent and mini-batches of size 128.

The model is trained for two epochs, keeping the pre-trained word vectors fixed in order to minimize large variations of the embeddings. Next, we allow the neural network to perform a fine-tuning of the word embeddings for the sentiment polarity classification task. The model is trained for 2 more epochs, projecting the pre-trained embeddings in a new space which takes into account information about the polarity of each word. This allows creating an embedding space where some regions are highly correlated with positive or negative opinions, thereby being more suitable for semantic sentiment analysis.

5   Results

The system has been evaluated using the word embeddings provided by the organizers of the challenge. During preliminary experiments, we used 60% of the dataset for training purposes, whereas 20% of the reviews have been used as a validation set and the remaining 20% as the test set. To produce the final results, however, we gave higher priority to the size of the training set, using 90% of the dataset for training the model and the remaining 10% as the validation set.

The best results have been obtained with the vectors of size $m = 256$, generated using 50 training epochs. Even without performing the fine-tuning process, in our experiments, these vectors allowed achieving a F-measure of 0.901 on a test set made up of the 20% of the original dataset. We have also experimented the fine-tuning of such embeddings, obtaining a *polarized* set of word vectors, that allowed scoring a F-measure of 0.956 on the same test set. The benefits of the fine-tuning process are even clearer when it is applied on different word embeddings, like the word vectors generated on the Google News dataset, as shown in Table 1.

Table 1. Comparison of the results obtained before and after performing the polarization of the embeddings

	Pre-trained	Polarized
Google News Embeddings	0.865	0.941
ESWC 2018 Embeddings	0.901	0.956

Our system has also performed very well in the ESWC 2018 Challenge, being ranked second (only 0.008 away from the winner) and obtaining a result comparable with what we measured during preliminary experiments. Table 2 lists the results of the challenge. An hyphen in F-measure values is present when the related systems provided results not accepted by the challenge chairs (e.g. wrong format, sent over the deadline).

Table 2. Results of ESWC 2018 Challenge on Semantic Sentiment Analysis (Task 1)

System	F-Measure
Guangyuan Piao and John G. Breslin **Domain-Aware Sentiment Classification with CNNs and LSTMs**	0.9643
Mattia Atzeni and Diego Reforgiato Recupero **Fine-Tuning of Word Embeddings for Semantic Sentiment Analysis**	0.9561
Marco Federici and Mauro Dragoni **Aspect-Based Opinion Mining Using Knowledge Bases**	0.9356
Mauro Dragoni **A Neural Word Embeddings Approach for Multi-Domain Sentiment Analysis**	0.9228
Mauro Dragoni **Fuzzy Logic for Aspect-based Sentiment Analysis**	0.8823
Andi Rexha **Exploiting Clauses for Opinion Mining**	0.8743
Giulio Petrucci **An Information Retrieval-based System For Multi-Domain Sentiment Analysis**	0.7153
Debanjan Chaudhuri and Jens Lehmann **CNN for Product Review Classification with Pre-Trained Word Embeddings**	0.5243
Vijayasaradhi Indurthi and Subba Reddy Oota **Evaluating Quality of Word Embeddings with Sentiment Polarity Identification Task**	–
Naveen Kumar Laskari and Suresh Kumar Sanampudi **Evaluation of Word Embedding Techniques and Sentiment Analysis Using LSTM**	–

As we can see, there is a significant improvement with respect to the previous run of the challenge [17], where the systems were heavily based on hand-crafted features. For instance, in the system introduced in [1] and better described in [6], we made use of a significant feature engineering process, aimed at combining syntactic and semantic features through both semantic replacement and augmentation. Table 3 shows the results of the challenge in 2017.

Table 3. Results of ESWC 2017 Challenge on Semantic Sentiment Analysis

System	F-Measure
Mattia Atzeni, Amna Dridi and Diego Reforgiato Recupero	0.8675
Fine-Grained Sentiment Analysis on Financial Microblogs and News Headlines [1]	
Marco Federici	0.8424
A Knowledge-based Approach For Aspect-Based Opinion Mining [7]	
Walid Iguider and Diego Reforgiato Recupero	0.8378
Language Independent Sentiment Analysis of the Shukran Social Network using Apache Spark [10]	
Giulio Petrucci	0.8112
The IRMUDOSA System at ESWC-2017 Challenge on Semantic Sentiment Analysis [14]	

6   Conclusion

In this paper we have shown a recent Deep Learning approach that has competed in the ESWC 2018 Challenge on Semantic Sentiment Analysis with promising results. The system is based on a single 2-layer Bidirectional LSTM, equipped with a neural attention mechanism. We have described the architecture of the network and have discussed its main advantages with respect to other models. Moreover, we experimented a *polarization* of the pre-trained word embeddings supplied by the organizers of the challenge. The evaluation of the approach provides evidence supporting the intuitions underlying the development of the system. Indeed, the system has revealed a high performance at the challenge and experimental results show the effectiveness of the *polarized* embeddings produced by the model.

Acknowledgements. The authors gratefully acknowledge Sardinia Regional Government for the financial support (Convenzione triennale tra la Fondazione di Sardegna e gli Atenei Sardi Regione Sardegna L.R. 7/2007 annualità 2016 DGR 28/21 del 17.05.2016, CUP: F72F16003030002). This work has been supported by Sardinia Regional Government (P.O.R. Sardegna F.S.E. Operational Programme of the Autonomous Region of Sardinia, European Social Fund 2014-2020 - Axis IV Human Resources, Objective 1.3, Line of Activity 1.3.1).

References

1. Atzeni, Mattia, Dridi, Amna, Reforgiato Recupero, Diego: Fine-grained sentiment analysis on financial microblogs and news headlines. In: Dragoni, Mauro, Solanki, Monika, Blomqvist, Eva (eds.) SemWebEval 2017. CCIS, vol. 769, pp. 124–128. Springer, Cham (2017). https://doi.org/10.1007/978-3-319-69146-6_11

2. Atzeni, Mattia, Reforgiato Recupero, Diego: Deep learning and sentiment analysis for human-robot interaction. In: Gangemi, Aldo, Gentile, Anna Lisa, Nuzzolese, Andrea Giovanni, Rudolph, Sebastian, Maleshkova, Maria, Paulheim, Heiko, Pan, Jeff Z., Alam, Mehwish (eds.) ESWC 2018. LNCS, vol. 11155, pp. 14–18. Springer, Cham (2018). https://doi.org/10.1007/978-3-319-98192-5_3

3. Baziotis, C., Pelekis, N., Doulkeridis, C.: DataStories at SemEval-2017 Task 4: deep LSTM with attention for message-level and topic-based sentiment analysis. In: Proceedings of the 11th International Workshop on Semantic Evaluation (SemEval-2017), pp. 747–754. Association for Computational Linguistics (2017)

4. Cliche, M.: BB_twtr at SemEval-2017 Task 4: Twitter sentiment analysis with CNNs and LSTMs. In: Proceedings of the 11th International Workshop on Semantic Evaluation (SemEval-2017). Association for Computational Linguistics (2017)

5. Collobert, R., Weston, J.: A unified architecture for natural language processing: deep neural networks with multitask learning. In: Cohen, W.W., McCallum, A., Roweis, S.T. (eds.) Proceedings of the Twenty-Fifth International Conference Machine Learning (ICML 2008), Helsinki, Finland, June 5–9, 2008. ACM International Conference Proceeding Series, vol. 307, pp. 160–167. ACM (2008). http://doi.acm.org/10.1145/1390156.1390177

6. Dridi, A., Atzeni, M., Reforgiato Recupero, D.: Finenews: fine-grained semantic sentiment analysis on financial microblogs and news. Int. J. Mach. Learn. Cybern. (2018). https://doi.org/10.1007/s13042-018-0805-x

7. Federici, M., Dragoni, M.: A knowledge-based approach for aspect-based opinion mining. In: Sack, H., Dietze, S., Tordai, A., Lange, C. (eds.) Semantic Web Challenges, pp. 141–152. Springer International Publishing, Cham (2016)

8. Gangemi, A., Presutti, V., Reforgiato Recupero, D.: Frame-based detection of opinion holders and topics: a model and a tool. IEEE Comput. Intell. Mag. 9(1), 20–30 (2014). https://www.scopus.com/inward/record.uri?eid=2-s2.0-84893375456&doi=10.1109%2fMCI.2013.2291688&partnerID=40&md5=d4133e755c2d02956702b55a9b8dc5ab

9. Hochreiter, S., Schmidhuber, J.: Long short-term memory. Neural Comput. 9(8), 1735–1780 (1997). Nov

10. Iguider, Walid, Reforgiato Recupero, Diego: Language independent sentiment analysis of the shukran social network using apache spark. In: Dragoni, Mauro, Solanki, Monika, Blomqvist, Eva (eds.) SemWebEval 2017. CCIS, vol. 769, pp. 129–132. Springer, Cham (2017). https://doi.org/10.1007/978-3-319-69146-6_12

11. Lecun, Y., Bottou, L., Bengio, Y., Haffner, P.: Gradient-based learning applied to document recognition. In: Proceedings of the IEEE. IEEE Press (1998)

12. Mikolov, T., Sutskever, I., Chen, K., Corrado, G.S., Dean, J.: Distributed representations of words and phrases and their compositionality. In: Burges, C.J.C., Bottou, L., Ghahramani, Z., Weinberger, K.Q. (eds.) Advances in Neural Information Processing Systems 26: 27th Annual Conference on Neural Information Processing Systems 2013. Proceedings of a meeting held December 5–8, 2013, Lake Tahoe, Nevada, United States, pp. 3111–3119. Curran Associates Inc. (2013). http://papers.nips.cc/paper/5021-distributed-representations-of-words-and-phrases-and-their-compositionality

13. Palogiannidi, E., et al.: Tweester at semeval-2016 task 4: sentiment analysis in twitter using semantic-affective model adaptation. In: Bethard, S., Cer, D.M., Carpuat, M., Jurgens, D., Nakov, P., Zesch, T. (eds.) Proceedings of the 10th International Workshop on Semantic Evaluation, SemEval@NAACL-HLT 2016. The Association for Computer Linguistics (2016)

14. Petrucci, Giulio, Dragoni, Mauro: The IRMUDOSA system at ESWC-2017 challenge on semantic sentiment analysis. In: Dragoni, Mauro, Solanki, Monika, Blomqvist, Eva (eds.) SemWebEval 2017. CCIS, vol. 769, pp. 148–165. Springer, Cham (2017). https://doi.org/10.1007/978-3-319-69146-6_14
15. Raffel, C., Ellis, D.P.W.: Feed-forward networks with attention can solve some long-term memory problems. CoRR abs/ arXiv:1512.08756 (2015). http://arxiv.org/abs/1512.08756
16. Recupero, D., Consoli, S., Gangemi, A., Nuzzolese, A., Spampinato, D.: A semantic web based core engine to efficiently perform sentiment analysis. Lecture Notes in Computer Science (including subseries Lecture Notes in Artificial Intelligence and Lecture Notes in Bioinformatics) vol. 8798, pp. 245–248 (2014). https://www.scopus.com/inward/record.uri?eid=2-s2.0-84908681970&doi=10.1007
17. Reforgiato Recupero, D., Cambria, E., Di Rosa, E.: Semantic sentiment analysis challenge at eswc2017. In: Dragoni, M., Solanki, M., Blomqvist, E. (eds.) Semantic Web Challenges, pp. 109–123. Springer International Publishing, Cham (2017)
18. Reforgiato Recupero, D., Presutti, V., Consoli, S., Gangemi, A., Nuzzolese, A.G.: Sentilo: frame-based sentiment analysis. Cognit. Comput. **7**(2), 211–225 (2015). https://doi.org/10.1007/s12559-014-9302-z
19. Rosenthal, S., Farra, N., Nakov, P.: SemEval-2017 task 4: sentiment analysis in Twitter. In: Proceedings of the 11th International Workshop on Semantic Evaluation. SemEval 2017. Association for Computational Linguistics (2017)
20. Srivastava, N., Hinton, G., Krizhevsky, A., Sutskever, I., Salakhutdinov, R.: Dropout: A simple way to prevent neural networks from overfitting. J. Mach. Learn. Res. **15**(1), 1929–1958 (2014). Jan
21. Stojanovski, D., Strezoski, G., Madjarov, G., Dimitrovski, I.: Finki at semeval-2016 task 4: deep learning architecture for twitter sentiment analysis. In: Proceedings of the 10th International Workshop on Semantic Evaluation (SemEval-2016), pp. 149–154. Association for Computational Linguistics (2016)

The KABSA System at ESWC-2018 Challenge on Semantic Sentiment Analysis

Marco Federici[1] and Mauro Dragoni[2]([⊠])

[1] Universitá di Trento, Trento, Italy
federici@fbk.eu
[2] Fondazione Bruno Kessler, Trento, Italy
dragoni@fbk.eu

Abstract. In the last decade, the focus of the Opinion Mining field moved to detection of the pairs "aspect-polarity" instead of limiting approaches in the computation of the general polarity of a text. In this work, we propose an aspect-based opinion mining system based on the use of semantic resources for the extraction of the aspects from a text and for the computation of their polarities. The proposed system participated at the third edition of the Semantic Sentiment Analysis (SSA) challenge took place during ESWC 2018 achieving the runner-up place in the Task #2 concerning the aspect-based sentiment analysis. Moreover, a further evaluation performed on the SemEval 2015 benchmarks demonstrated the feasibility of the proposed approach.

1 Introduction And Related Work

Opinion Mining is a natural language processing (NLP) task that aims to classify documents according to their opinion (polarity) on a given subject [1]. This task has created a considerable interest due to its wide applications in different domains: marketing, politics, social sciences, etc.. Generally, the polarity of a document is computed by analyzing the expressions contained in the full text without distinguishing which are the subjects of each opinion. In the last decade, the research in the opinion mining field focused on the "aspect-based opinion mining" [2] consisting in the extraction of all subjects ("aspects") from documents and the opinions that are associated with them.

For clarification, let us consider the following example:

Yesterday, I bought a new smartphone.
The quality of the display is very good, but the buttery lasts too little.

In the sentence above, we may identify three aspects: "smartphone", "display", and "battery". As the reader may see, each aspect has a different opinion associated with it. The list below summarizes such associations:

D. Buscaldi et al. (Eds.): SemWebEval 2018, CCIS 927, pp. 151–166, 2018.
https://doi.org/10.1007/978-3-030-00072-1_13

- "display" → "very good"
- "battery" → "too little"
- "smarthphone" → no explicit opinions, therefore polarity can be inferred by averaging the opinions associated with all other aspects.

The topic of aspect-based sentiment analysis has been explored under different perspectives. A comprehensive review of the last available systems can be found in the proceedings of SemEval 2015[1].

The paper is structured as follows. Section 2 briefly provides an overview of the aspect extraction task. Section 3 introduces the background knowledge used during the development of the system. Section 4 presents the underlying NLP layer upon which it has been developed the system described Sect. 5. Sections 6 and 7 shows the results obtained on the ESWC 2016 SSA challenge and on the SemEval 2015 benchmark, respectively. Finally, Sect. 8 provide a description about how the tasks of the challenge have been addressed and it concludes the paper.

2   Related Work

The topic of sentiment analysis has been studied extensively in the literature [3], where several techniques have been proposed and validated.

Machine learning techniques are the most common approaches used for addressing this problem, given that any existing supervised methods can be applied to sentiment classification. For instance, in [4], the authors compared the performance of Naive-Bayes, Maximum Entropy, and Support Vector Machines in sentiment analysis on different features like considering only unigrams, bigrams, combination of both, incorporating parts of speech and position information or by taking only adjectives. Moreover, beside the use of standard machine learning method, researchers have also proposed several custom techniques specifically for sentiment classification, like the use of adapted score function based on the evaluation of positive or negative words in product reviews [5], as well as by defining weighting schemata for enhancing classification accuracy [6].

An obstacle to research in this direction is the need of labeled training data, whose preparation is a time-consuming activity. Therefore, in order to reduce the labeling effort, opinion words have been used for training procedures. In [7] and [8], the authors used opinion words to label portions of informative examples for training the classifiers. Opinion words have been exploited also for improving the accuracy of sentiment classification, as presented in [9], where a framework incorporating lexical knowledge in supervised learning to enhance accuracy has been proposed. Opinion words have been used also for unsupervised learning approaches like the one presented in [10].

Another research direction concerns the exploitation of discourse-analysis techniques. [11] discusses some discourse-based supervised and unsupervised

---

[1] http://alt.qcri.org/semeval2015/cdrom/pdf/SemEval082.pdf.

approaches for opinion analysis; while in [12], the authors present an approach to identify discourse relations.

The approaches presented above are applied at the document-level[13–18], i.e., the polarity value is assigned to the entire document content. However, in some case, for improving the accuracy of the sentiment classification, a more fine-grained analysis of a document is needed. Hence, the sentiment classification of the single sentences, has to be performed. In the literature, we may find approaches ranging from the use of fuzzy logic [19–23] to the use of aggregation techniques [24] for computing the score aggregation of opinion words. In the case of sentence-level sentiment classification, two different sub-tasks have to be addressed: (i) to determine if the sentence is subjective or objective, and (ii) in the case that the sentence is subjective, to determine if the opinion expressed in the sentence is positive, negative, or neutral. The task of classifying a sentence as subjective or objective, called "subjectivity classification", has been widely discussed in the literature [25–28] and systems implementing the capabilities of identifying opinion's holder, target, and polarity have been presented [29]. Once subjective sentences are identified, the same methods as for sentiment classification may be applied. For example, in [30] the authors consider gradable adjectives for sentiment spotting; while in [31–33] the authors built models to identify some specific types of opinions.

In the last years, with the growth of product reviews, the use of sentiment analysis techniques was the perfect floor for validating them in marketing activities [34–36]. However, the issue of improving the ability of detecting the different opinions concerning the same product expressed in the same review became a challenging problem. Such a task has been faced by introducing "aspect" extraction approaches that were able to extract, from each sentence, which is the aspect the opinion refers to. In the literature, many approaches have been proposed: conditional random fields (CRF) [37], hidden Markov models (HMM) [38], sequential rule mining [39], dependency tree kernels [40], clustering [41], neural networks [42,43], and genetic algorithms [44]. In [45,46], two methods were proposed to extract both opinion words and aspects simultaneously by exploiting some syntactic relations of opinion words and aspects.

A particular attention should be given also to the application of sentiment analysis in social networks [47–49]. More and more often, people use social networks for expressing their moods concerning their last purchase or, in general, about new products. Such a social network environment opened up new challenges due to the different ways people express their opinions, as described by [50] and [51], who mention "noisy data" as one of the biggest hurdles in analyzing social network texts.

One of the first studies on sentiment analysis on micro-blogging websites has been discussed in [52], where the authors present a distant supervision-based approach for sentiment classification.

At the same time, the social dimension of the Web opens up the opportunity to combine computer science and social sciences to better recognize, interpret, and process opinions and sentiments expressed over it. Such multi-disciplinary approach has been called *sentic computing* [53]. Application domains where sentic computing has already shown its potential are the cognitive-inspired classification of images [54], of texts in natural language, and of handwritten text [55].

Finally, an interesting recent research direction is domain adaptation, as it has been shown that sentiment classification is highly sensitive to the domain from which the training data is extracted. A classifier trained using opinionated documents from one domain often performs poorly when it is applied or tested on opinionated documents from another domain, as we demonstrated through the example presented in Sect. 1. The reason is that words and even language constructs used in different domains for expressing opinions can be quite different. To make matters worse, the same word in one domain may have positive connotations, but in another domain may have negative ones; therefore, domain adaptation is needed. In the literature, different approaches related to the Multi-Domain sentiment analysis have been proposed. Briefly, two main categories may be identified: (i) the transfer of learned classifiers across different domains [56–58], and (ii) the use of propagation of labels through graph structures [19,59–61].

All approaches presented above are based on the use of statistical techniques for building sentiment models. The exploitation of semantic information is not taken into account. In this work, we proposed a first version of a semantic-based approach preserving the semantic relationships between the terms of each sentence in order to exploit them either for building the model and for estimating document polarity. The proposed approach, falling into the multi-domain sentiment analysis category, instead of using pre-determined polarity information associated with terms, it learns them directly from domain-specific documents. Such documents are used for training the models used by the system.

3   Preliminaries

The system is implemented on top of a background knowledge used for representing the linguistic connections between "concepts" described in several resources. Below, it is possible to find the list of such resources and the links where further information about them may be found.

WordNet[2] [62] is one of the most important resource available to researchers in the field of text analysis, computational linguistics, and many related areas. In the implemented system, WordNet has been used as starting point for the construction of the semantic graph used by the system (see Sect. 5) However, due to some coverage limitations occurring in WordNet, it has been extended by linking further terms coming from the Roget's Thesaurus [63].

SenticNet[3] [64] is a publicly available resource for opinion mining that exploits both Artificial Intelligence and Semantic Web techniques to infer the polarity associated with common-sense concepts and represent it in a semantic-aware format. In particular, SenticNet uses dimensionality reduction to calcu-

---

[2] https://wordnet.princeton.edu/.
[3] http://sentic.net/.

late the affective valence of a set of Open Mind concepts and represent it in a machine-accessible and machine-processable format.

General Inquirer dictionary[4] [65] is an English-language dictionary containing almost 12,000 elements associated with their polarity in different contexts. Such dictionary is the result of the integration between the "Harvard" and the "Lasswell" general-purpose dictionaries as well as a dictionary of categories define by the dictionary creators. When necessary, for ambiguous words, specific polarity for each sense is specified.

4   The Underlying NLP Layer

The presented system has been implemented on top of existing Natural Language Processing libraries. In particular, it uses different functionalities offered by the Stanford NLP Library.

WordNet[5] [62] resource is used together with Stanford's part of speech annotation to detect compound nouns. Lists of consecutive nouns and word sequences contained in Wordnet compound nouns vocabulary are merged into a single word in order to force Stanford library to consider them as a single unit during the following phases. The entire text is then fed to the co-reference resolution module to compute pronoun references which are stored in an index-reference map. Details about the textual analysis are provided in Sect. 5.

The next operation consists in detecting which word expresses polarity within each sentence. To achieve this task *SenticNet*, *General Inquirer dictionary* and *MPQA* sentiment lexicons have been used.

While SenticNet expresses polarity values in the continuous range from -1 to 1, the other two resources been normalized: the General Inquirer words have positive values of polarity if they belong to the "Positive" class while negative if they belong to "Negative" one, zero otherwise, similarly, MPQA "polarity" labels are used to infer a numerical values. Only words with a non-zero polarity value in at least one resource are considered as opinion words (e.g. word "third" is not present in MPQA and SenticNet and has a 0 value according to General Inquirer, consequently, it is not a valid opinion word; on the other hand, word "huge" has a positive 0.069 value according to SenticNet, a negative value in MPQA and 0 value according to General Inquirer, therefore, it is a possible opinion word even if lexicons express contrasting values). Every noun (single or complex) is considered an aspect as long as it's connected to at least one opinion and it's not in the stopword list. This list has been created starting from the "Onix" text retrieval engine stopwords list[6] and it contains words without a specific meaning (such as "thing") and special characters.

Opinions associated with pronouns are connected to the aspect they are referring to; instead, if pronouns reference can't be resolved, they are both discarded.

---

[4] http://www.wjh.harvard.edu/~inquirer/spreadsheet_guide.htm.

[5] https://wordnet.princeton.edu/.

[6] The used stopwords list is available at http://www.lextek.com/manuals/onix/stopwords1.html.

The main task of the system is, then, represented by connecting opinions with possible aspects. Two different approaches have been tested with a few variants. The first one relies on the syntactic tree while the second one is based on grammar dependencies.

The sentence "I enjoyed the screen resolution, it's amazing for such a cheap laptop." has been used to underline differences in connection techniques.

The preliminary phase merges words "screen" and "resolution" into a single word "Screenresolution" because they are consecutive nouns. Co-reference resolution module extracts a relation between "it" and "Screenresolution". This relation is stored so that every possible opinion that would be connected to "it" will be connected to "Screenresolution" instead. Figure 1 shows the syntax tree while Fig. 2 represents the grammar relation graph generated starting from the example sentence. Both structures have been computed using Stanford NLP modules ("parse", "depparse").

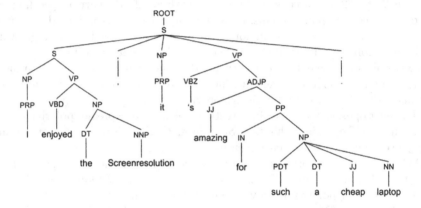

Fig. 1. Example of syntax tree.

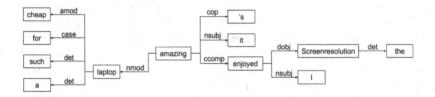

Fig. 2. Example of the grammar relations graph.

5   The Implemented System

The aspect extraction component is based on a six-phases approach as presented below.

Phase 1 Sentences are given as input to the Stanford NLP Library[7] and they are annotated with part of speech (POS) tags in order to detect nouns, adjectives, and pronouns.

Phase 2 Tokens annotated as adjectives are considered for computing opinion scores, while sequences of one or more consecutive nouns (for example "support" tagged as "NN" followed by "team" tagged "NN" as well) and complex linguistic structures recognized through the use of Wordnet (for example "hard" annotated as "JJ" and "disk" annotated as "NN") are aggregated and marked as potential aspects. This step is shown in Fig. 3

Phase 3 Co-reference resolution is applied for resolving pronouns co-references between nouns. Example about how co-reference is applied is shown in Fig. 3.

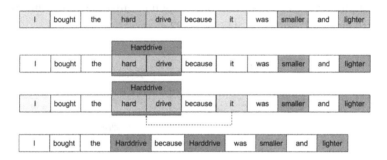

Fig. 3. Example of noun aggregation and co-reference resolution.

Phase 4 After the aggregation of compound names, we changed the sentences by replacing compound names with single tokens to ensure that they are considered as single entities during the opinion resolution phase. This way, it will be possible to exchange each pronoun with the corresponding label of the aspect they are referring to.

Phase 5 Stanford Parser is used for generating a syntax tree that is exploited in the last phase for associating opinions with aspects. Concerning the definition of the associations between aspects and opinions, during preliminary testing activities, we tried different approaches. Among them:

- each aspect has been connected with each opinion contained in the same sentence, where as sentence delimiters, we used the markers "S", "SBAR", and "FRAG" detected in the parsed tree;
- if an opinion is expressed in a sentence without nouns, such an opinion has been associated with the aspects belonging to the same noun phrase only.

Example of the generated parsed tree is shown in Fig. 4.

---
[7] http://stanfordnlp.github.io/CoreNLP/index.html.

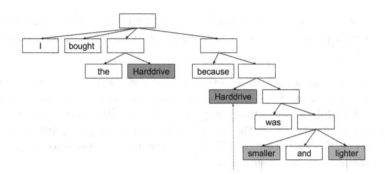

Fig. 4. Example of generated parse tree.

Phase 6 Finally, aspects without associated opinions are discarded, while remaining ones are stored. Example is shown in Fig. 5.

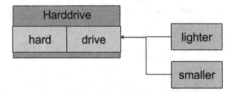

Fig. 5. Example of opinion association.

6   ESWC-2018 SSA Challenge Tasks #1 and #2

The expected output of this task was a set of aspects of the reviewed product and a binary polarity value associated to each of such aspects. So, for example, while for classic binary polarity inference task an overall polarity (positive or negative) was expected for a review about a mobile phone, this task required a set of aspects (such as *speaker, touchscreen, camera*, etc.) and a polarity value (positive or negative) associated with each of such aspects. Engines were assessed according to both aspect extraction and aspect polarity detection using precision, recall, f-measure, and accuracy similarly as performed during the first edition of the Concept-Level Sentiment Analysis challenge held during ESWC2014 and re-proposed at SemEval 2015 Task 12[8]. Please refer to SemEval 2016 Task 5[9] for details on the precision-recall analysis. By considering that the aspect-based sentiment analysis includes also the necessity of inferring concept polarities, we decided to apply our system also to Task #1. Figures 6 and 7 show an example of the output schema for Tasks #1 and #2.

---

[8] http://www.alt.qcri.org/semeval2015/task12/.
[9] http://alt.qcri.org/semeval2016/task5/.

```
<?xml version="1.0" encoding="UTF-8" standalone="yes"?>
<Sentences>
    <sentence id="apparel_0">
        <text>
            GOOD LOOKING KICKS IF YOUR KICKIN IT OLD SCHOOL LIKE ME. AND COMFORTABLE.
            AND RELATIVELY CHEAP. I'LL ALWAYS KEEP A PAIR OF STAN SMITH'S
            AROUND FOR WEEKENDS
        </text>
        <polarity>
        positive
        </polarity>
    </sentence>
    <sentence id="apparel_1">
        <text>
            These sunglasses are all right. They were a little crooked, but still cool..
        </text>
        <polarity>
        positive
        </polarity>
    </sentence>
```

Fig. 6. Task #1 output example. Input is the same without the opinion tag and its descendant nodes.

```
<?xml version="1.0" encoding="UTF-8" standalone="yes"?>
<Review rid="1">
    <sentences>
        <sentence id="348:0">
            <text>Most everything is fine with this machine: speed, capacity, build.</text>
            <Opinions>
                <Opinion aspect="MACHINE" polarity="positive"/>
            </Opinions>
        </sentence>
        <sentence id="348:1">
            <text>The only thing I don't understand is that the resolution of the
            screen isn't high enough for some pages, such as Yahoo!Mail.
            </text>
            <Opinions>
                <Opinion aspect="SCREEN" polarity="negative"/>
            </Opinions>
        </sentence>
        <sentence id="277:2">
            <text>The screen takes some getting use to, because it is smaller
            than the laptop.</text>
            <Opinions>
                <Opinion aspect="SCREEN" polarity="negative"/>
            </Opinions>
        </sentence>
    </sentences>
</Review>
```

Fig. 7. Task #2 output example. Input is the same without the opinion tag and its descendant nodes.

The training set was composed by 5,058 sentences coming from two different domains: "Laptop" (3,048 sentences) and "Restaurant" (2,000 sentences). While, the test set was composed by 891 sentences coming from the "Laptop" (728 sentences) and "Hotels" (163 sentence). The reason for which we decided to use the "Hotels" domain in the test set with respect to the "Restaurant" one was to observe the capability of the participant systems to be general purpose with respect to the training set.

7 System Evaluation and Error Analysis

The system has been tested on two aspect-based sentiment analysis datasets by following the "Semi-Open" setting of the DRANZIERA protocol [66]:

D1 The SemEval 2015 Task 12 training set benchmark, consisting in sentences belonging to the "Laptop" and "Restaurant" domains.
D2 The ESWC2018 Benchmark on Semantic Sentiment Analysis test set, consisting in sentences belonging to the "Laptop" and "Hotels" domains.

To compute results, a notion of correctness has to be introduced: if the extracted aspects is equal, contained or contains the correct one, it's considered to be correct (for example if the extracted aspect is "screen", while the annotated one is "screen of the computer" or vice versa, the result of the system is considered to correct).

Tables 1 shows the results obtained on each dataset; while, Tables 2 and 3 shows the full results of Tasks #1 and #2 participants.

Table 1. Results obtained by the presented system on the SemEval 2015 Task 12 dataset and on the test set adopted for the challenge.

Dataset	Precision	Recall	F-Measure	Polarity accuracy
D1	0.39969	0.39478	0,39722	0.92170
D2	0.34820	0.35745	0.35276	0.85284

Figure 8 shows an analysis of error cases. Values have been computed according to the first 100 sentences of the "Laptop" dataset.

The majority of false negatives are given by the impossibility to detect opinions expressed by verbs. For example, in the sentence "I generally like this place" or more complex expressions "tech support would not fix the problem unless I bought your plan for $150 plus".

Other issues are correlated to the association algorithm. Figure 9 shows the specific error analysis related to the extraction of aspects, always computed on the same 100 sentences of the "Laptop" dataset.

Even if the syntax-tree-based approach tends to produce a significant number of true positives, relationships are often imprecise. A relevant example is represented by the sentence "I was extremely happy with the OS itself." in the "Laptop" dataset. The approach connects the opinion adjective "happy" with the potential aspect "OS", correctly recognized as an aspect in the sentence.

A relevant part of false positives are generated due to the incapability of discriminating aspects from the entity itself. In facts, almost half of them consists

Table 2. Results obtained by the participant to Task #1.

System	F-Mesure
Guangyuan et. al	0.9643
Atzori et. al	0.9561
Federici et. al	**0.9356**
Dragoni et. al	0.9228
Dragoni et. al	0.8823
Rexha et. al	0.8743
Indurthi et. al	0.8203
Petrucci et. al	0.7153
Chaudhuri et. al	0.5243

Table 3. Results obtained by the participant to Task #2.

System	F-Mesure
Federici et. al	**0.5682**
Dragoni et. al	0.5244
Rexha et. al	0.4598

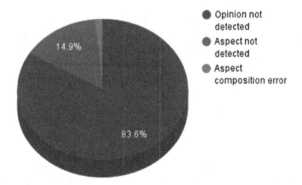

Fig. 8. Overall error analysis.

in associations between opinion words and the entity reviewed that are correct. However, they must not be considered during the aspect extraction task (for example the aspect "laptop" in the example sentence should not be considered according to the definition of aspect).

8   Conclusions

In this paper, we presented the system submitted to the third edition of the Semantic Sentiment Analysis run during ESWC 2016. The system participated

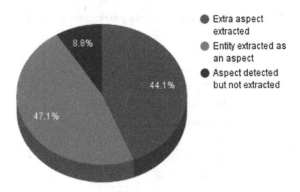

Fig. 9. Specific error analysis on aspect extraction.

only at Task #2 and obtained the second place. The problem of detecting aspects in sentences is very relevant in the sentiment analysis community. Further work in this direction will be performed by starting from the analysis of the errors provided in the previous section.

References

1. Pang, B., Lee, L., Vaithyanathan, S.: Thumbs up? sentiment classification using machine learning techniques. In: Proceedings of EMNLP, Philadelphia, pp. 79–86. Association for Computational Linguistics (July 2002)
2. Hu, M., Liu, B.: Mining and summarizing customer reviews. In: Proceedings of the Tenth ACM SIGKDD International Conference on Knowledge Discovery and Data Mining, pp. 168–177. ACM (2004)
3. Liu, B., Zhang, L.: A survey of opinion mining and sentiment analysis. In: Aggarwal, C.C., Zhai, C.X. (eds.) Mining Text Data, pp. 415–463. Springer (2012)
4. Pang, B., Lee, L.: A sentimental education: Sentiment analysis using subjectivity summarization based on minimum cuts. In: ACL, pp. 271–278 (2004)
5. Dave, K., Lawrence, S., Pennock, D.M.: Mining the peanut gallery: opinion extraction and semantic classification of product reviews. In: WWW, pp. 519–528 (2003)
6. Paltoglou, G., Thelwall, M.: A study of information retrieval weighting schemes for sentiment analysis. In: ACL, pp. 1386–1395 (2010)
7. Tan, S., Wang, Y., Cheng, X.: Combining learn-based and lexicon-based techniques for sentiment detection without using labeled examples. In: SIGIR, pp. 743–744 (2008)
8. Qiu, L., Zhang, W., Hu, C., Zhao, K.: Selc: a self-supervised model for sentiment classification. In: CIKM, pp. 929–936 (2009)
9. Melville, P., Gryc, W., Lawrence, R.D.: Sentiment analysis of blogs by combining lexical knowledge with text classification. In: KDD, pp. 1275–1284 (2009)
10. Taboada, M., Brooke, J., Tofiloski, M., Voll, K.D., Stede, M.: Lexicon-based methods for sentiment analysis. Comput. Linguist. **37**(2), 267–307 (2011)
11. Somasundaran, S.: Discourse-level relations for Opinion Analysis. Ph.D. thesis, University of Pittsburgh (2010)

12. Wang, H., Zhou, G.: Topic-driven multi-document summarization. In: IALP, pp. 195–198 (2010)
13. Dragoni, M.: Shellfbk: An information retrieval-based system for multi-domain sentiment analysis. In: Proceedings of the 9th International Workshop on Semantic Evaluation. SemEval '2015, Denver, Colorado, pp. 502–509. Association for Computational Linguistics (June 2015)
14. Petrucci, G., Dragoni, M.: An information retrieval-based system for multi-domain sentiment analysis. In: Gandon, F., Cabrio, E., Stankovic, M., Zimmermann, A. (eds.) SemWebEval 2015. CCIS, vol. 548, pp. 234–243. Springer, Cham (2015). https://doi.org/10.1007/978-3-319-25518-7_20
15. Rexha, A., Kröll, M., Dragoni, M., Kern, R.: Exploiting propositions for opinion mining. In: Sack, H., Dietze, S., Tordai, A., Lange, C. (eds.) SemWebEval 2016. CCIS, vol. 641, pp. 121–125. Springer, Cham (2016). https://doi.org/10.1007/978-3-319-46565-4_9
16. Federici, M., Dragoni, M.: A knowledge-based approach for aspect-based opinion mining. In: Sack, H., Dietze, S., Tordai, A., Lange, C. (eds.) SemWebEval 2016. CCIS, vol. 641, pp. 141–152. Springer, Cham (2016). https://doi.org/10.1007/978-3-319-46565-4_11
17. Rexha, A., Kröll, M., Dragoni, M., Kern, R.: Opinion mining with a clause-based approach. In: Dragoni, M., Solanki, M., Blomqvist, E. (eds.) SemWebEval 2017. CCIS, vol. 769, pp. 166–175. Springer, Cham (2017). https://doi.org/10.1007/978-3-319-69146-6_15
18. Federici, M., Dragoni, M.: Aspect-based opinion mining using knowledge bases. In: Dragoni, M., Solanki, M., Blomqvist, E. (eds.) SemWebEval 2017. CCIS, vol. 769, pp. 133–147. Springer, Cham (2017). https://doi.org/10.1007/978-3-319-69146-6_13
19. Dragoni, M., Tettamanzi, A.G., da Costa Pereira, C.: Propagating and aggregating fuzzy polarities for concept-level sentiment analysis. Cognit. Comput. **7**(2), 186–197 (2015)
20. Dragoni, M., Tettamanzi, A.G.B., da Costa Pereira, C.: A fuzzy system for concept-level sentiment analysis. In: Presutti, V., Stankovic, M., Cambria, E., Cantador, I., Di Iorio, A., Di Noia, T., Lange, C., Reforgiato Recupero, D., Tordai, A. (eds.) SemWebEval 2014. CCIS, vol. 475, pp. 21–27. Springer, Cham (2014). https://doi.org/10.1007/978-3-319-12024-9_2
21. Petrucci, G., Dragoni, M.: The IRMUDOSA system at ESWC-2016 challenge on semantic sentiment analysis. In: Sack, H., Dietze, S., Tordai, A., Lange, C. (eds.) SemWebEval 2016. CCIS, vol. 641, pp. 126–140. Springer, Cham (2016). https://doi.org/10.1007/978-3-319-46565-4_10
22. Dragoni, M., Petrucci, G.: A fuzzy-based strategy for multi-domain sentiment analysis. Int. J. Approx. Reason. **93**, 59–73 (2018)
23. Petrucci, G., Dragoni, M.: The IRMUDOSA system at ESWC-2017 challenge on semantic sentiment analysis. In: Dragoni, M., Solanki, M., Blomqvist, E. (eds.) SemWebEval 2017. CCIS, vol. 769, pp. 148–165. Springer, Cham (2017). https://doi.org/10.1007/978-3-319-69146-6_14
24. da Costa Pereira, C., Dragoni, M., Pasi, G.: A prioritized "and" aggregation operator for multidimensional relevance assessment. In Serra, R., Cucchiara, R. (eds.) AI*IA 2009: Emergent Perspectives in Artificial Intelligence, XIth International Conference of the Italian Association for Artificial Intelligence, Reggio Emilia, Italy, December 9–12, 2009, Proceedings. Volume 5883 of Lecture Notes in Computer Science, pp. 72–81. Springer (2009)

25. Sack, H., Rizzo, G., Steinmetz, N., Mladenić, D., Auer, S., Lange, C. (eds.): ESWC 2016. LNCS, vol. 9989. Springer, Cham (2016). https://doi.org/10.1007/978-3-319-47602-5

26. Federici, M., Dragoni, M.: A branching strategy for unsupervised aspect-based sentimentanalysis. In Dragoni, M., Recupero, D.R. (eds.) Proceedings of the 3rd International Workshop at ESWC on Emotions, Modality, Sentiment Analysis and the SemanticWeb co-located with 14th ESWC 2017, Portroz, Slovenia, May 28, 2017, CEUR Workshop Proceedings, vol. 1874 (2017). http://ceur-ws.org/

27. Riloff, E., Patwardhan, S., Wiebe, J.: Feature subsumption for opinion analysis. In: EMNLP, pp. 440–448 (2006)

28. Wilson, T., Wiebe, J., Hwa, R.: Recognizing strong and weak opinion clauses. Comput. Intell. **22**(2), 73–99 (2006)

29. Palmero Aprosio, A., Corcoglioniti, F., Dragoni, M., Rospocher, M.: Supervised opinion frames detection with RAID. In: Gandon, F., Cabrio, E., Stankovic, M., Zimmermann, A. (eds.) SemWebEval 2015. CCIS, vol. 548, pp. 251–263. Springer, Cham (2015). https://doi.org/10.1007/978-3-319-25518-7_22

30. Hatzivassiloglou, V., Wiebe, J.: Effects of adjective orientation and gradability on sentence subjectivity. In: COLING, pp. 299–305 (2000)

31. Kim, S.M., Hovy, E.H.: Crystal: analyzing predictive opinions on the web. In: EMNLP-CoNLL, pp. 1056–1064 (2007)

32. Rexha, A., Kröll, M., Dragoni, M., Kern, R.: Polarity classification for target phrases in tweets: a Word2Vec approach. In: Sack, H., Rizzo, G., Steinmetz, N., Mladenić, D., Auer, S., Lange, C. (eds.) ESWC 2016. LNCS, vol. 9989, pp. 217–223. Springer, Cham (2016). https://doi.org/10.1007/978-3-319-47602-5_40

33. Blomqvist, E., Maynard, D., Gangemi, A., Hoekstra, R., Hitzler, P., Hartig, O. (eds.): ESWC 2017. LNCS, vol. 10250. Springer, Cham (2017). https://doi.org/10.1007/978-3-319-58451-5

34. Recupero, D.R., Dragoni, M., Presutti, V.: ESWC 15 challenge on concept-level sentiment analysis. In: Gandon, F., Cabrio, E., Stankovic, M., Zimmermann, A. (eds.) SemWebEval 2015. CCIS, vol. 548, pp. 211–222. Springer, Cham (2015). https://doi.org/10.1007/978-3-319-25518-7_18

35. Dragoni, M., Reforgiato Recupero, D.: Challenge on fine-Ggrained sentiment analysis within ESWC2016. In: Sack, H., Dietze, S., Tordai, A., Lange, C. (eds.) SemWebEval 2016. CCIS, vol. 641, pp. 79–94. Springer, Cham (2016). https://doi.org/10.1007/978-3-319-46565-4_6

36. Dragoni, M., Solanki, M., Blomqvist, E. (eds.): SemWebEval 2017. CCIS, vol. 769. Springer, Cham (2017). https://doi.org/10.1007/978-3-319-69146-6

37. Jakob, N., Gurevych, I.: Extracting opinion targets in a single and cross-domain setting with conditional random fields. In: EMNLP, pp. 1035–1045 (2010)

38. Jin, W., Ho, H.H., Srihari, R.K.: Opinionminer: a novel machine learning system for web opinion mining and extraction. In: KDD, pp. 1195–1204 (2009)

39. Liu, B., Hu, M., Cheng, J.: Opinion observer: analyzing and comparing opinions on the web. In: WWW, pp. 342–351 (2005)

40. Wu, Y., Zhang, Q., Huang, X., Wu, L.: Phrase dependency parsing for opinion mining. In: EMNLP, pp. 1533–1541 (2009)

41. Su, Q., et al.: Hidden sentiment association in chinese web opinion mining. In: WWW, pp. 959–968 (2008)

42. Dragoni, M.: NEUROSENT-PDI at semeval-2018 task 1: leveraging a multi-domain sentiment model for inferring polarity in micro-blog text. In: Apidianaki, M., Mohammad, S.M., May, J., Shutova, E., Bethard, S., Carpuat, M. (eds.) Proceedings of The 12th International Workshop on Semantic Evaluation, SemEval@NAACL-HLT, New Orleans, Louisiana, 5–6 June 2018, pp. 102–108. Association for Computational Linguistics (2018)

43. Dragoni, M.: NEUROSENT-PDI at semeval-2018 task 3: Understanding irony in social networks through a multi-domain sentiment model. In: Apidianaki, M., Mohammad, S.M., May, J., Shutova, E., Bethard, S., Carpuat, M. (eds.) Proceedings of The 12th International Workshop on Semantic Evaluation, SemEval@NAACL-HLT, New Orleans, Louisiana, 5–6 June 2018, pp. 512–519. Association for Computational Linguistics (2018)

44. Dragoni, M., Azzini, A., Tettamanzi, A.G.B.: A novel similarity-based crossover for artificial neural network evolution. In: Schaefer, R., Cotta, C., Kołodziej, J., Rudolph, G. (eds.) PPSN 2010. LNCS, vol. 6238, pp. 344–353. Springer, Heidelberg (2010). https://doi.org/10.1007/978-3-642-15844-5_35

45. Qiu, G., Liu, B., Bu, J., Chen, C.: Opinion word expansion and target extraction through double propagation. Comput. Linguist. **37**(1), 9–27 (2011)

46. Dragoni, M., da Costa Pereira, C., Tettamanzi, A.G.B., Villata, S.: Combining argumentation and aspect-based opinion mining: the smack system. AI Commun. **31**(1), 75–95 (2018)

47. Dragoni, M.: A three-phase approach for exploiting opinion mining in computational advertising. IEEE Intell. Syst. **32**(3), 21–27 (2017)

48. Dragoni, M., Petrucci, G.: A neural word embeddings approach for multi-domain sentiment analysis. IEEE Trans. Affect. Comput. **8**(4), 457–470 (2017)

49. Dragoni, M.: Computational advertising in social networks: an opinion mining-based approach. In: Haddad, H.M., Wainwright, R.L., Chbeir, R. (eds.) Proceedings of the 33rd Annual ACM Symposium on Applied Computing, SAC 2018, Pau, France, 09–13 April 2018, pp. 1798–1804. ACM (2018)

50. Barbosa, L., Feng, J.: Robust sentiment detection on twitter from biased and noisy data. In: COLING (Posters), pp. 36–44. (2010)

51. Bermingham, A., Smeaton, A.F.: Classifying sentiment in microblogs: is brevity an advantage? In: CIKM, pp. 1833–1836 (2010)

52. Go, A., Bhayani, R., Huang, L.: Twitter sentiment classification using distant supervision. CS224N Project Report, Standford University (2009)

53. Cambria, E., Hussain, A.: Sentic Computing: A Common-Sense-Based Framework For Concept-Level Sentiment Analysis (2015)

54. Cambria, E., Hussain, A.: Sentic album: content-, concept-, and context-based online personal photo management system. Cognit. Comput. **4**(4), 477–496 (2012)

55. Wang, Q.F., Cambria, E., Liu, C.L., Hussain, A.: Common sense knowledge for handwritten chinese recognition. Cognit. Comput. **5**(2), 234–242 (2013)

56. Blitzer, J., Dredze, M., Pereira, F.: Biographies, bollywood, boom-boxes and blenders: domain adaptation for sentiment classification. In: ACL, pp. 187–205 (2007)

57. Pan, S.J., Ni, X., Sun, J.T., Yang, Q., Chen, Z.: Cross-domain sentiment classification via spectral feature alignment. In: WWW, pp. 751–760 (2010)

58. Yoshida, Y., Hirao, T., Iwata, T., Nagata, M., Matsumoto, Y.: Transfer learning for multiple-domain sentiment analysis–identifying domain dependent/independent word polarity. AAA I, 1286–1291 (2011)

59. Ponomareva, N., Thelwall, M.: Semi-supervised vs. cross-domain graphs for sentiment analysis. In: RANLP, pp. 571–578 (2013)

60. Huang, S., Niu, Z., Shi, C.: Automatic construction of domain-specific sentiment lexicon based on constrained label propagation. Knowl.-Based Syst. **56**, 191–200 (2014)
61. Dragoni, M., da Costa Pereira, C., Tettamanzi, A.G.B., Villata, S.: Smack: An argumentation framework for opinion mining. In: Kambhampati, S., ed.: Proceedings of the Twenty-Fifth International Joint Conference on Artificial Intelligence, IJCAI 2016, New York, NY, USA, 9–15 July 2016, IJCAI/AAAI Press, pp. 4242–4243 (2016)
62. Fellbaum, C.: WordNet: An Electronic Lexical Database. MIT Press, Cambridge (1998)
63. Kipfer, B.A.: Roget's 21st century thesaurus, 3rd edn (2005)
64. Cambria, E., Speer, R., Havasi, C., Hussain, A.: Senticnet: a publicly available semantic resource for opinion mining. In: AAAI Fall Symposium: Commonsense Knowledge (2010)
65. P.J., S., Dunphy, D., Marshall, S.: The General Inquirer: A Computer Approach to Content Analysis. MIT Press, Oxford (1966)
66. Dragoni, M., Tettamanzi, A., da Costa Pereira, C.: Dranziera: an evaluation protocol for multi-domain opinion mining. In: Chair, N.C.C., et al. (eds.) Proceedings of the Tenth International Conference on Language Resources and Evaluation (LREC 2016), Paris, France, European Language Resources Association (ELRA) (May, 2016)

The IRMUDOSA System at ESWC-2018 Challenge on Semantic Sentiment Analysis

Giulio Petrucci[1] and Mauro Dragoni[2]($\boxtimes$)

[1] Universitá di Trento, Trento, Italy
petrucci@fbk.eu
[2] Fondazione Bruno Kessler, Trento, Italy
dragoni@fbk.eu

Abstract. Multi-domain opinion mining consists in estimating the polarity of a document by exploiting domain-specific information. One of the main issue of the approaches discussed in literature is their poor capability of being applied on domains that have not been used for building the opinion model. In this paper, we present an approach exploiting the linguistic overlap between domains for building models enabling the estimation of polarities for documents belonging to any other domain. The system implementing such an approach has been presented at the third edition of the Semantic Sentiment Analysis Challenge co-located with ESWC 2018. Fuzzy representation of features polarity supports the modeling of information uncertainty learned from training set and integrated with knowledge extracted from two well-known resources used in the opinion mining field, namely Sentic.Net and the General Inquirer. The proposed technique has been validated on a multi-domain dataset and the results demonstrated the effectiveness of the proposed approach by setting a plausible starting point for future work.

1 Introduction

Opinion mining is a natural language processing task aiming to classify documents according to the opinion (polarity) they express on a given subject [1]. Generally speaking, opinion mining aims at determining the attitude of a speaker or a writer with respect to a topic or the overall tonality of a document. This task has created a considerable interest due to its wide applications. In recent years, the exponential increase of the Web for exchanging public opinions about events, facts, products, etc., has led to an extensive usage of opinion mining approaches, especially for marketing purposes.

Most of the work available in the literature address the opinion mining problem without distinguishing the domains which documents, used for building models, come from. The necessity of investigating this problem from a multi-domain perspective is led by the different influence that a term might have in different contexts. Let us consider the following examples. In the first example, we have

© Springer Nature Switzerland AG 2018
D. Buscaldi et al. (Eds.): SemWebEval 2018, CCIS 927, pp. 167–185, 2018.
https://doi.org/10.1007/978-3-030-00072-1_14

an "emotion-based" context where the adjective "cold" is used differently based on the feeling, or mood, of the opinion holder:

1. That person always behaves in a very **cold** way with her colleagues.
2. A **cold** drink is the best thing we can drink when the temperature is very hot.

while in the second one, we have a "subjective-based" context where the adjective "small" is used differently based on the product category reviewed by a user:

1. The sideboard is **small** and it is not able to contain a lot of stuff.
2. The **small** dimensions of this decoder allow to move it easily.

In the first context, we considered two different "emotional" situations: in the first one a person is commenting about the behavior of his colleague by using the adjective "cold" with a "negative" polarity. Instead, in the second one, a person is referring to the adjective "cold" in a "positive" way as a good solution for a situation.

Instead, in the second context, we considered the interpretation of texts referring to two different domain: "Furnishings" and "Electronics". In the first one, the polarity of the adjective "small" is, for sure, negative because it highlights an issue of the described item. On the other hand, in the second domain, the polarity of such an adjective may be considered positive.

The multiple facets with which textual information can be analyzed in the context of opinion mining led to the design of approaches creating models able to address this scenario. The idea of adapting terms polarity to different domains emerged only recently [2]. In general, multi-domain opining mining approaches discussed in the literature (surveyed in Sect. 2) focus on building models for transferring information between pairs of domains [3]. While on one hand such approaches allow to propagate specific domain information to other, their drawback is the necessity of building new transfer models any time a new domain has to be addressed. This way, approaches include a poor generalization capability of analyzing text, because transfer models are limited to the N domains used for building the models.

This paper describes our approach exploiting the linguistic overlap between domains for building models enabling the estimation of polarities for documents. Due to this peculiarity, the proposed approach is innovative, to the best of our knowledge, with respect to the state of the art of multi-domain opinion mining.

The rest of the article is structured as follows. Section 2 presents a survey on works about opinion mining either in the single o multi domain environment. Section 3 provides the references to the knowledge resources used in the implementation of the proposed approach described in detail in Sect. 4. Section 5 reports the system evaluation and, finally, Sect. 6 concludes the article.

2   Related Work

The topic of sentiment analysis has been studied extensively in the literature [4], where several techniques have been proposed and validated.

Machine learning techniques are the most common approaches used for addressing this problem, given that any existing supervised methods can be applied to sentiment classification. For instance, in [5], the authors compared the performance of Naive-Bayes, Maximum Entropy, and Support Vector Machines in sentiment analysis on different features like considering only unigrams, bigrams, combination of both, incorporating parts of speech and position information or by taking only adjectives. Moreover, beside the use of standard machine learning method, researchers have also proposed several custom techniques specifically for sentiment classification, like the use of adapted score function based on the evaluation of positive or negative words in product reviews [6], as well as by defining weighting schemata for enhancing classification accuracy [7].

An obstacle to research in this direction is the need of labeled training data, whose preparation is a time-consuming activity. Therefore, in order to reduce the labeling effort, opinion words have been used for training procedures. In [8] and [9], the authors used opinion words to label portions of informative examples for training the classifiers. Opinion words have been exploited also for improving the accuracy of sentiment classification, as presented in [10], where a framework incorporating lexical knowledge in supervised learning to enhance accuracy has been proposed. Opinion words have been used also for unsupervised learning approaches like the one presented in [11].

Another research direction concerns the exploitation of discourse-analysis techniques. [12] discusses some discourse-based supervised and unsupervised approaches for opinion analysis; while in [13], the authors present an approach to identify discourse relations.

The approaches presented above are applied at the document-level [14–19], i.e., the polarity value is assigned to the entire document content. However, in some case, for improving the accuracy of the sentiment classification, a more fine-grained analysis of a document is needed. Hence, the sentiment classification of the single sentences, has to be performed. In the literature, we may find approaches ranging from the use of fuzzy logic [20–24] to the use of aggregation techniques [25] for computing the score aggregation of opinion words. In the case of sentence-level sentiment classification, two different sub-tasks have to be addressed: (i) to determine if the sentence is subjective or objective, and (ii) in the case that the sentence is subjective, to determine if the opinion expressed in the sentence is positive, negative, or neutral. The task of classifying a sentence as subjective or objective, called "subjectivity classification", has been widely discussed in the literature [26–29] and systems implementing the capabilities of identifying opinion's holder, target, and polarity have been presented [30]. Once subjective sentences are identified, the same methods as for sentiment classification may be applied. For example, in [31] the authors consider gradable adjectives for sentiment spotting; while in [32–34] the authors built models to identify some specific types of opinions.

In the last years, with the growth of product reviews, the use of sentiment analysis techniques was the perfect floor for validating them in marketing activities [35–37]. However, the issue of improving the ability of detecting the different opinions concerning the same product expressed in the same review became a challenging problem. Such a task has been faced by introducing "aspect" extraction approaches that were able to extract, from each sentence, which is the aspect the opinion refers to. In the literature, many approaches have been proposed: conditional random fields (CRF) [38], hidden Markov models (HMM) [39], sequential rule mining [40], dependency tree kernels [41], clustering [42], neural networks [43,44], and genetic algorithms [45]. In [46,47], two methods were proposed to extract both opinion words and aspects simultaneously by exploiting some syntactic relations of opinion words and aspects.

A particular attention should be given also to the application of sentiment analysis in social networks [48–50]. More and more often, people use social networks for expressing their moods concerning their last purchase or, in general, about new products. Such a social network environment opened up new challenges due to the different ways people express their opinions, as described by [51,52], who mention "noisy data" as one of the biggest hurdles in analyzing social network texts.

One of the first studies on sentiment analysis on micro-blogging websites has been discussed in [53], where the authors present a distant supervision-based approach for sentiment classification.

At the same time, the social dimension of the Web opens up the opportunity to combine computer science and social sciences to better recognize, interpret, and process opinions and sentiments expressed over it. Such multi-disciplinary approach has been called *sentic computing* [54]. Application domains where sentic computing has already shown its potential are the cognitive-inspired classification of images [55], of texts in natural language, and of handwritten text [56].

Finally, an interesting recent research direction is domain adaptation, as it has been shown that sentiment classification is highly sensitive to the domain from which the training data is extracted. A classifier trained using opinionated documents from one domain often performs poorly when it is applied or tested on opinionated documents from another domain, as we demonstrated through the example presented in Sect. 1. The reason is that words and even language constructs used in different domains for expressing opinions can be quite different. To make matters worse, the same word in one domain may have positive connotations, but in another domain may have negative ones; therefore, domain adaptation is needed. In the literature, different approaches related to the Multi-Domain sentiment analysis have been proposed. Briefly, two main categories may be identified: (i) the transfer of learned classifiers across different domains [2,3,57], and (ii) the use of propagation of labels through graph structures [20,58–60].

All approaches presented above are based on the use of statistical techniques for building sentiment models. The exploitation of semantic information is not taken into account. In this work, we proposed a first version of a semantic-based approach preserving the semantic relationships between the terms of each sentence in order to exploit them either for building the model and for estimating document polarity. The proposed approach, falling into the multi-domain sentiment analysis category, instead of using pre-determined polarity information associated with terms, it learns them directly from domain-specific documents. Such documents are used for training the models used by the system.

3 Knowledge Resources

The proposed approach exploits the use of background knowledge for supporting the creation of the multi-domain model used for computing text polarities. Such a background knowledge is composed by two linguistic resources freely available to the research community. Below, we briefly describe them, while in Sect. 4, we present how they have been used for supporting the implementation of the proposed approach.

SenticNet SenticNet[1] [61] is a publicly available resource for opinion mining that exploits both artificial intelligence and semantic Web techniques to infer the polarities associated with common-sense concepts and to represent them in a semantic-aware format. In particular, SenticNet uses dimensionality reduction to calculate the affective valence of a set of Open Mind[2] concepts and it represents them in a machine accessible and processable format. SenticNet contains more than 5,700 polarity concepts (nearly 40% of the Open Mind corpus) and it may be connected with any kind of opinion mining application. For example, after the de-construction of the text into concepts through a semantic parser, SenticNet can be used to associate polarity values to these and, hence, infer the overall polarity of a clause, sentence, paragraph, or document by averaging such values.

General Inquirer General Inquirer dictionary [3] [62] is an English-language dictionary containing almost 12,000 elements associated with their polarity in different contexts. Such dictionary is the result of the integration between the "Harvard" and the "Lasswell" general-purpose dictionaries as well as a dictionary of categories define by the dictionary creators. When necessary, for ambiguous words, specific polarity for each sense is specified. For every words, a set of tags is provided in the dictionary. Among them, only a subset is relevant to the opinion mining topic and have been exploited in this work: "valence categories", "semantic dimensions", "words of pleasure", and "words reflecting presence or lack of emotional expressiveness". Other categories indicating ascriptive social categories rather than references to places have been considered out of the scope of the opinion mining topic and have not been considered in the implementation of the approach.

4 Method

The main goal of the presented approach is to exploit domain overlap for compensating the lack of knowledge caused by building models using only a snap-

---

[1] http://sentic.net/.

[2] http://commons.media.mit.edu/en/.

[3] http://www.wjh.harvard.edu/~inquirer/spreadsheet_guide.htm.

shot of the reality. When a domain-based model is built, part of the knowledge belonging to such a domain is not included in the model due to its missing in the adopted training set. Besides, when a system has to classify a text belonging to a domain that has not been used for building the model, it is possible that such a model does not contain enough information for estimating text opinion. For this reason, it is necessary to compensate this lack of knowledge by partially exploiting information coming from other domains.

In this section, we describe the steps adopted for building the models and the strategy we implemented for computing the polarity of a text by exploiting domain overlapping.

The model construction process is composed by three steps: (i) features are extracted from each text contained in the training set (Sect. 4.1); then, (ii) a preliminary fuzzy membership functions [63], modeled by using a triangular shape, is computed by analyzing only the explicit information contained in the dataset (Sect. 4.2); and, (iii) this shape is transformed into a trapezoid after a refinement operation performed by compensating the uncertainty inherited by the adoption of a training set, with information coming from external resources: the SenticNet knowledge base and the General Inquirer dictionary (Sect. 4.3).

Finally, when the inference of a text polarity is requested, the usage of the model for estimating text polarities is performed by aggregating the polarities associated with each feature detected from the text to polarize and by computing the final judgment (Sect. 4.4).

4.1   Feature Extraction

During the Feature Extraction (FE) phase, documents are analyzed and significant elements are extracted and used as features for building the model. As feature, we mean every text chunk that may have a meaning in the context of opinion mining and/or in domain detection. The first step that we performed for extracting the features is to parse the content of each document by using the Stanford NLP Parser [64]. The parser has been used for annotating terms with part of speech (POS) tags and for extracting the dependencies tree of each sentence.

Let's consider the following text marked with a positive polarity: "This smartphone is great. The display is awesome and the touch system works very well."

By parser the text, we obtained the analysis of the two detected sentences. Their POS-tagged versions are represented in the lines 1 and 2; while, the dependencies of the first sentence are shown in the lines from 3 to 6 and, finally, the dependencies of the second sentence are shown in the lines from 7 to 17.

```
1.  This/DT smartphone/NN is/VBZ great/JJ ./.
2.  The/DT display/NN is/VBZ awesome/JJ and/CC the/DT
    touch/NN system/NN works/VBZ very/RB well/RB ./.

3.  root ( ROOT-0 , great-4 )
4.  det ( smartphone-2 , This-1 )
```

```
 5.  nsubj ( great-4 , smartphone-2 )
 6.  cop ( great-4 , is-3 )

 7.  root ( ROOT-0 , awesome-4 )
 8.  det ( display-2 , The-1 )
 9.  nsubj ( awesome-4 , display-2 )
10.  cop ( awesome-4 , is-3 )
11.  cc ( awesome-4 , and-5 )
12.  det ( system-8 , the-6 )
13.  compound ( system-8 , touch-7 )
14.  nsubj ( works-9 , system-8 )
15.  conj:and ( awesome-4 , works-9 )
16.  advmod ( well-11 , very-10 )
17.  advmod ( works-9 , well-11 )
```

From the parser output, we distinguished two type of features for building our model:

– Single concepts feature: nouns, adjectives, and verbs are stored in the model as single features. Nouns are used for building the domain detection component of our model; while, adjectives and verbs are used for building the opinion mining component. By considering as example the dependencies extracted from the first sentence, the term "smartphone" is inserted in the model with the role of supporting the domain detection, while the term "great" is inserted in the model with the role of supporting the definition about how positive polarity is modeled.
– Terms dependency feature: a selection of the dependencies extracted by the parser is stored within the model with the aim of incorporating domain specific contextual knowledge describing (i) how concepts are connected in a particular domain and (ii) how such connections are related to a particular polarity. The kind of dependencies took into account are "noun-adjective", "adjective-verb", "noun-verb", and "adjective-adverb". In the example above, lines containing significant terms dependency features are 5, 9, 10, 14, 16, and 17. From each "term dependency" feature we actually extract further features that are inserted in the model as well. Let's consider as example the dependency at line 16. From the dependency, we extract "well-very", "very-well", "well", and "very".

4.2 Preliminary Learning Phase

The Preliminary Learning (PL) phase aims at estimating the starting polarity and the domain belonging degree (DBD) of each feature. The estimation of these values is done by analyzing only the explicit information provided by the training set.

Concerning the estimation of the feature polarity, this phase allows to define the preliminary fuzzy membership functions representing the polarity of each feature extracted from the training set with respect to the domain containing such polarity. The feature polarity is estimated as:

$$p_i^E(F) = \frac{k_F^i}{T_F^i} \in [-1, 1] \qquad \forall i = 1, \ldots, n, \tag{1}$$

where F is the feature taken into account, index i refers to domain D_i which the feature belongs to, n is the number of domains available in the training set, k_F^i is the arithmetic sum of the polarities observed for feature F in the training set restricted to domain D_i, T_F^i is the number of instances of the training set, restricted to domain D_i, in which feature F occurs, and E stays for "estimated". The shape of the fuzzy membership function generated during this phase is a triangle with the top vertex in the coordinates $(x, 1)$, where $x = p_i^{(E)}(F)$ and with the two bottom vertexes in the coordinates $(-1, 0)$ and $(1, 0)$ respectively. The rationale is that while we have one point (x) in which we have full confidence, our uncertainty covers the entire space because we do not have any information concerning the remaining polarity values. At this stage, the types of feature took into account are the terms dependency features and, as single concept features, adjectives and verbs.

Figure 1 shows a picture of the generated fuzzy triangle.

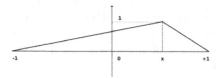

Fig. 1. The fuzzy triangle generated after the Preliminary Learning Phase.

After the polarity estimation, we computed the DBD of each feature. Such a value is exploited during the Polarity Aggregation and Decision Phase (described in Sect. 4.4) for computing the final polarity of a document.

The computation of the DBD is inspired by the well-known TF-IDF model [65] used in information retrieval, where the importance of a term is given by either the frequency of a term in a document contained in an index and the inverse of the number of documents in which such a term occurs.

In our case, the DBD of a feature is computed by summing two factors: the feature frequency associated with the domain in which the feature occurs and the "uniqueness" of the feature with respect to all domains. The domain-frequency is computed as:

$$\text{freq}_i(F) = \frac{k_i(F)}{N_i} \tag{2}$$

where F is the feature taken into account, i is the domain that is analyzed, k_i is the number of times that the feature F occurs in the domain i, and N_i is the total number of features contained in the domain i.

While, the feature uniqueness is computed as:

$$\text{uniq}_i(F) = \frac{\text{freq}_i(F)}{\sum\limits_{i=0}^{n} \text{freq}_i(F)} \tag{3}$$

where F is the feature taken into account and n is the number of domains.
Finally the DBD of each feature is given by:

$$\text{DBD}_i(F) = \text{freq}_i(F) + \text{uniq}_i(F) \tag{4}$$

4.3   Information Refinement Phase

Polarities estimated during the PL phase are refined by exploiting, for each
feature, polarities extracted from the resources described in Sect. 3. The rationale
behind this choice is to balance the polarity estimated from the training set
(that represents only a snapshot of the world) with polarity information that
are contained in supervised knowledge bases. When we estimate the polarity
value of each feature, two scenarios may happen:

1. the estimated polarity "agrees" (i.e. it has the same orientation) with the one
   extracted from the knowledge bases;
2. the estimated polarity "disagrees" with the one extracted from the knowledge
   bases.

In the first case, the estimated polarity confirms, in terms of opinion orientation, what it is represented in the knowledge bases. The representation of this
kind of uncertainty will be a tight shape.

On the contrary, in the second case, the estimated polarity is the opposite of
what has been extracted from the knowledge bases. In this case, the uncertainty
associated with the feature will produce a larger shape. Such a shape will model
the contrast between what has been estimated from the training set and what
has been defined by experts in the construction of the knowledge bases.

For this reason an Information Refinement (IR) phase is necessary in order
to convert this uncertainty in a numerical representation that can be managed
by the system.

Assume to have the following values associated to the feature F belonging
to the domain i:

- p_s^F, represents the polarity of the feature F extracted from SenticNet;
- p_g^F, represents the polarity of the feature F extracted from the General
  Inquirer;
- avg_p is the average polarity computed among $p_i^E(F)$, p_s^F, and p_g^F;
- var_p is the variance computed between avg_p and the three polarities values
  $p_i^E(F)$, p_s^F, and p_g^F.

For "terms dependency" features, that are composed by two terms T_1 and T_2, the values p_s^F and p_g^F are the average of the single polarities computed on T_1 and T_2, respectively.

By starting from these values, the final shape of the inferred fuzzy membership functions, at the end of the IR phase, is a trapezoid whose core consists of the interval between the polarity value learned during the PL phase, $p_i^E(F)$, and avg_p. While, the support of the fuzzy shape is given, on both sides, by the variance var_p.

To sum up, for each domain D_i, $\mu_{F,i}$ is a trapezoid with parameters (a, b, c, d), where

$$a = \min\{p_i^E(F), avg_p\},$$
$$b = \max\{p_i^E(F), avg_p\},$$
$$c = \max\{-1, a - var_p\},$$
$$d = \min\{1, b + var_p\}.$$

The idea here is that the most likely values for the polarity of F for domain D_i are those comprised between the estimated value and average between the estimation of the training algorithm and the polarity values retrieved from supervised knowledge resources. The uncertainty modeled by the fuzzy shape is proportional to the level of "agreement" between the estimated polarity value, and the polarities retrieved from the supervised knowledge bases.

Figure 2 shows a picture of the generated fuzzy trapezoid.

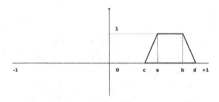

Fig. 2. The fuzzy trapezoid generated after the Information Refinement Phase.

4.4   Polarity Aggregation and Decision Phase

The fuzzy polarities of different features, resulting from the IR phase, are finally aggregated by a fuzzy averaging operator obtained by applying the extension principle [66] in order to compute fuzzy polarities for complex entities, like texts, which consist of a number of features and thus derive, so to speak, their polarity from them. When a crisp polarity value is needed, it may be computed from a fuzzy polarity by applying a defuzzification [67] method.

The operation of computing the entire polarity of a text is done not only on the model describing the domain which the document belongs, but also on the other domains. Indeed, one of the assumption of the proposed approach is to exploit possible linguistic overlaps for compensating missing knowledge of the training set. Therefore, given $p_i(D)$ as the polarity computed on document D with respect to the model of the domain i, and $DBD_i(D)$ as the domain belonging degree of document D to the domain i, we compute the following two vectors:

$$\langle \text{polarity}_D \rangle = [p_0(D), p_1(D), \ldots, p_n(D)] \tag{5}$$

and

$$\langle \text{domain}_D \rangle = [DBD_0(D), DBD_1(D), \ldots, DBD_n(D)] \tag{6}$$

where n is the number of domains contained in the model. The final polarity of a text is then computed by multiplying the two vectors as follow:

$$p_T = \frac{\langle \text{polarity}_D \rangle \times \langle \text{domain}_D \rangle}{N} \tag{7}$$

where N is the number of domains used for building the model.

5   System Evaluation

Here, we present the evaluation procedure adopted for validating the proposed approach.

5.1   The Dataset

The system has been trained by using the DRANZIERA dataset [68]. The dataset is composed by one million reviews crawled from product pages on the Amazon web site[4]. Such reviews belong to twenty different domains, we called in-model domains (IMD): Amazon Instant Video, Automotive, Baby, Beauty, Books, Clothing Accessories, Electronics, Health, Home Kitchen, Movies TV, Music, Office Products, Patio, Pet Supplies, Shoes, Software, Sports Outdoors, Tools Home Improvement, Toys Games, and Video Games.

For each domain, we extracted twenty-five thousands positive and twenty-five thousands negative reviews that have been split in five folds containing five thousand positive and five thousand negative reviews each. This way, the dataset is balanced with respect to either the polarities of the reviews and to the domain which they belong to. The choice between positive and negative documents has been inspired by the strategy used in [2] where reviews with 4 or 5 stars have been marked as positive, while the ones having 1 or 2 stars have been marked as negative.

Besides the twenty domains mentioned above, we used further 7 test sets for measuring the effectiveness of the approach in estimating polarities of texts

---

[4] All the material used for the evaluation and the built models are available at http://goo.gl/pj0nWS.

belonging to domains different from the ones used to build the model, we called out-model domains (OMD). Such domains are: Cell Phones Accessories, Gourmet Foods, Industrial Scientific, Jewelry, Kindle Store, Musical Instruments, and Watches.

5.2   Evaluation Procedure

The approach has been evaluated through a 5-cross-fold evaluation procedure. For each execution we measured the precision and the recall and, at the end, we report their averages together with the standard deviation measured over the five executions.

The approach has been compared with three baselines:

– Most Frequent Polarity (MFP): results obtained by guessing always the same polarity for all instances contained in the test set.
– Domain Belonging Polarity (DBP): results obtained by computing the text polarity by using only the information of the domain the text belongs to. This means that the linguistic overlap between domains has not been considered.
– Domain Detection Polarity (DDP): results obtained by computing the text polarity by using only the information of the domain guessed as the most appropriate one for the text to evaluate. This means that the similarity between text content and domain is preferred with respect to the domain used for tagging the text.

The same baselines have been used for evaluating the OMD test sets. In this case, the DBP baseline has not been applied due to the mismatch between the domains used for building the model and the ones contained in the test sets. Each OMD test set has been applied to all five models built, and the scores averaged.

5.3   In-Vitro Results

Here, we show the results of the evaluation campaign conducted for validating the presented approach. Tables 1 and 2 present a summary of the performance obtained by our system and by the three baselines on IMD and OMD, respectively. First column contains the name of the approach, second, third and fourth ones contain the average precision, recall and F1 score computed over all domains; while, the fifth column contains the average standard deviation computed on the F1 score during the cross-fold validation. Finally, the sixth and seventh columns contain the minimum and the maximum F1 score measured during the evaluation.

Table 1 shows the results obtained on in-domain models; while in Table 2, we present the results obtained by testing our approach on out-model domains.

In Table 2, the results about the DBP baseline are not reported because the used test set contains texts belonging to domains that are not included in the model.

Table 1. Comparison between the results obtained by the three baselines and the ones obtained by the proposed system on in-model domains.

Approach	Avg. precision	Avg. recall	Avg. F1
MFP	0.5000	1.0000	0.6667
DBP	0.7218	0.9931	0.8352
DDP	0.7115	0.9946	0.8290
DAP	**0.7686**	**0.9984**	**0.8679**
Approach	Avg. deviation	Avg. min. F1	Avg. max. F1
MFP	0.0000	0.6667	0.6667
DBP	0.5028	0.8153	0.8543
DDP	0.5584	0.8121	0.8456
DAP	**0.5954**	**0.8469**	**0.8881**

Table 2. Comparison between the results obtained by the two baselines and the ones obtained by the proposed system on out-model domains.

Approach	Avg. precision	Avg. recall	Avg. F1
MFP	0.5000	1.0000	0.6667
DBP	-	-	-
DDP	0.6766	0.9931	0.8045
DAP	**0.7508**	**0.9985**	**0.8564**
Approach	Avg. deviation	Avg. min. F1	Avg. max. F1
MFP	0.0000	0.6667	0.6667
DDP	0.5796	0.7906	0.8198
DAP	**0.6060**	**0.8389**	**0.8755**

By considering the overall results obtained on the IMD (Table 1), we may observe how the proposed approach outperforms the provided baselines. The measured F1 scores (average, minimum, and maximum) are higher of about 4% with respect to the baselines. The same happens for the Precision value, while for the Recall, all systems are very close to 100% and now significant differences have been detected. Test instances for which polarity has not been estimated have been judged as "neutral". Concerning the stability of the approach, we can notice that the exploitation of the domain information leads to a lower deviation over the five folds.

The second overall evaluation concerns the analysis of the results obtained on the set of OMD. The interesting aspect of this evaluation is to measure how the system is able to address the task of detecting polarities of documents coming from a different set of domains with respect to the ones used to build the model. Results are shown in Table 2. The first thing that we may observe is how the effectiveness obtained by the proposed system is very close to the one obtained in the IMD evaluation. Indeed, the difference between the two F1 averages is only around 1%. This aspect remarked the capability of the proposed approach to work in a cross-domain environment and to exploit the linguistic overlaps between domains for estimating text polarities. In this second evaluation, it is also possible to notice how the DDP baseline obtained lower results with respect to the evaluation on the IMD. This result confirms that a solution based on exploiting information coming from a domain resulting the "most similar" to the text to analyze are inadequate for computing text polarities.

In light of these results, we may state that the exploitation of linguistic overlaps between domains is a suitable solution for compensating the possible lack of knowledge had by building opinion models on limited training sets.

5.4 The IRMUDOSA System at ESWC-2018 SSA Challenge Task #1

The system participated to the Task #1 of the Semantic Sentiment Analysis Challenge co-located with ESWC 2016. Table 3 shows the results of Task #1. We may observe that the system ranked third, not far from the best performer. This result confirm the viability of the implemented approach for future implementation after a study of the main error scenarios in which the fuzzy-based algorithm performed poorly.

Table 3. Results obtained by the participant to Task #1.

System	F-mesure
Guangyuan et al.	0.9643
Atzori et al.	0.9561
Federici et al.	0.9356
Dragoni et al.	0.9228
Dragoni et al.	0.8823
Rexha et al.	0.8743
Indurthi et al.	0.8203
Petrucci et al.	**0.7153**
Chaudhuri et al.	0.5243

6   Conclusion and Future Work

In this article, we presented an approach to multi-domain opinion mining exploiting linguistic overlaps between domains for estimating the polarity of texts. The approach is supported by the implementation of a fuzzy model used for representing either the polarity of each feature with respect to a particular domain and its associated uncertainty.

Models are built by combining information extracted from a training set with the knowledge contained in two supervised linguistic resources, Sentic.net and the General Inquirer. The estimation of polarities is performed by combining the degree which a text belongs to each domain with each domain-specific polarity information extracted from the model.

Results shown the effectiveness improvement of the proposed approach with respect to the baselines demonstrating its viability and the close gap between the proposed system and the best performer participated to the Task #1 of Semantic Sentiment Analysis Challenge proved the potential of the fuzzy-based solution. Moreover, the protocol used for the evaluation enables an easy reproducibility of the experiments and the comparison of obtained results with other systems.

Future work will focus either on the enrichment of the knowledge used for building the models and on the use of fuzzy membership functions. Finally, we foresee the integration of a concept extraction approach in order to equip the system with further semantic capabilities of extracting finer-grained information (i.e., single aspects and semantic information associated with them) which can be used during the model construction.

References

1. Pang, B., Lee, L., Vaithyanathan, S.: Thumbs up? Sentiment classification using machine learning techniques. In: Proceedings of EMNLP, Philadelphia, Association for Computational Linguistics, pp. 79–86 (2002)
2. Blitzer, J., Dredze, M., Pereira, F.: Biographies, bollywood, boom-boxes and blenders: domain adaptation for sentiment classification. In: ACL, pp. 187–205 (2007)
3. Pan, S.J., Ni, X., Sun, J.T., Yang, Q., Chen, Z.: Cross-domain sentiment classification via spectral feature alignment. In: WWW, pp. 751–760 (2010)
4. Liu, B., Zhang, L.: A survey of opinion mining and sentiment analysis. In: Aggarwal, C.C., Zhai, C.X. (eds.) Mining Text Data, pp. 415–463. Springer (2012)
5. Pang, B., Lee, L.: A sentimental education: sentiment analysis using subjectivity summarization based on minimum cuts. In: ACL, pp. 271–278 (2004)
6. Dave, K., Lawrence, S., Pennock, D.M.: Mining the peanut gallery: opinion extraction and semantic classification of product reviews. In: WWW, pp. 519–528 (2003)
7. Paltoglou, G., Thelwall, M.: A study of information retrieval weighting schemes for sentiment analysis. In: ACL, pp. 1386–1395 (2010)
8. Tan, S., Wang, Y., Cheng, X.: Combining learn-based and lexicon-based techniques for sentiment detection without using labeled examples. In: SIGIR, pp. 743–744 (2008)
9. Qiu, L., Zhang, W., Hu, C., Zhao, K.: SELC: a self-supervised model for sentiment classification. In: CIKM, pp. 929–936 (2009)
10. Melville, P., Gryc, W., Lawrence, R.D.: Sentiment analysis of blogs by combining lexical knowledge with text classification. In: KDD, pp. 1275–1284 (2009)
11. Taboada, M., Brooke, J., Tofiloski, M., Voll, K.D., Stede, M.: Lexicon-based methods for sentiment analysis. Comput. Linguist. **37**(2), 267–307 (2011)
12. Somasundaran, S.: Discourse-level relations for opinion analysis. PhD thesis, University of Pittsburgh (2010)

13. Wang, H., Zhou, G.: Topic-driven multi-document summarization. In: IALP, pp. 195–198 (2010)
14. Dragoni, M.: Shellfbk: an information retrieval-based system for multi-domain sentiment analysis. In: Proceedings of the 9th International Workshop on Semantic Evaluation. SemEval '2015, Denver, Colorado, Association for Computational Linguistics, pp. 502–509 (2015)
15. Petrucci, G., Dragoni, M.: An information retrieval-based system for multi-domain sentiment analysis. In: Gandon, F., Cabrio, E., Stankovic, M., Zimmermann, A. (eds.) SemWebEval 2015. CCIS, vol. 548, pp. 234–243. Springer, Cham (2015). https://doi.org/10.1007/978-3-319-25518-7_20
16. Rexha, A., Kröll, M., Dragoni, M., Kern, R.: Exploiting propositions for opinion mining. In: Sack, H., Dietze, S., Tordai, A., Lange, C. (eds.) SemWebEval 2016. CCIS, vol. 641, pp. 121–125. Springer, Cham (2016). https://doi.org/10.1007/978-3-319-46565-4_9
17. Federici, M., Dragoni, M.: A knowledge-based approach for aspect-based opinion mining. In: Sack, H., Dietze, S., Tordai, A., Lange, C. (eds.) SemWebEval 2016. CCIS, vol. 641, pp. 141–152. Springer, Cham (2016). https://doi.org/10.1007/978-3-319-46565-4_11
18. Rexha, A., Kröll, M., Dragoni, M., Kern, R.: Opinion mining with a clause-based approach. In: Dragoni, M., Solanki, M., Blomqvist, E. (eds.) SemWebEval 2017. CCIS, vol. 769, pp. 166–175. Springer, Cham (2017). https://doi.org/10.1007/978-3-319-69146-6_15
19. Federici, M., Dragoni, M.: Aspect-based opinion mining using knowledge bases. In: Dragoni, M., Solanki, M., Blomqvist, E. (eds.) SemWebEval 2017. CCIS, vol. 769, pp. 133–147. Springer, Cham (2017). https://doi.org/10.1007/978-3-319-69146-6_13
20. Dragoni, M., Tettamanzi, A.G., da Costa Pereira, C.: Propagating and aggregating fuzzy polarities for concept-level sentiment analysis. Cogn. Comput. $7(2)$, 186–197 (2015)
21. Dragoni, M., Tettamanzi, A.G.B., da Costa Pereira, C.: A fuzzy system for concept-level sentiment analysis. In: Presutti, V., Stankovic, M., Cambria, E., Cantador, I., Di Iorio, A., Di Noia, T., Lange, C., Reforgiato Recupero, D., Tordai, A. (eds.) SemWebEval 2014. CCIS, vol. 475, pp. 21–27. Springer, Cham (2014). https://doi.org/10.1007/978-3-319-12024-9_2
22. Petrucci, G., Dragoni, M.: The IRMUDOSA system at ESWC-2016 challenge on semantic sentiment analysis. In: Sack, H., Dietze, S., Tordai, A., Lange, C. (eds.) SemWebEval 2016. CCIS, vol. 641, pp. 126–140. Springer, Cham (2016). https://doi.org/10.1007/978-3-319-46565-4_10
23. Dragoni, M., Petrucci, G.: A fuzzy-based strategy for multi-domain sentiment analysis. Int. J. Approx. Reason. 93, 59–73 (2018)
24. Petrucci, G., Dragoni, M.: The IRMUDOSA system at ESWC-2017 challenge on semantic sentiment analysis. In: Dragoni, M., Solanki, M., Blomqvist, E. (eds.) SemWebEval 2017. CCIS, vol. 769, pp. 148–165. Springer, Cham (2017). https://doi.org/10.1007/978-3-319-69146-6_14
25. da Costa Pereira, C., Dragoni, M., Pasi, G.: A prioritized "and" aggregation operator for multidimensional relevance assessment. In: Serra, R., Cucchiara, R. (eds.) AI*IA 2009: Emergent Perspectives in Artificial Intelligence, XIth International Conference of the Italian Association for Artificial Intelligence, Reggio Emilia, Italy, December 9–12, 2009, Proceedings. Volume 5883 of Lecture Notes in Computer Science, pp. 72–81. Springer (2009)

26. Federici, M., Dragoni, M.: Towards unsupervised approaches for aspects extraction. In: Dragoni, M., Recupero, D.R., Denecke, K., Deng, Y., Declerck, T. (eds.) Joint Proceedings of the 2th Workshop on Emotions, Modality, Sentiment Analysis and the Semantic Web and the 1st International Workshop on Extraction and Processing of Rich Semantics from Medical Texts co-located with ESWC 2016, Heraklion, Greece, May 29, 2016. Volume 1613 of CEUR Workshop Proceedings., CEUR-WS.org (2016)
27. Federici, M., Dragoni, M.: A branching strategy for unsupervised aspect-based sentiment analysis. In: Dragoni, M., Recupero, D.R. (eds.) Proceedings of the 3rd International Workshop at ESWC on Emotions, Modality, Sentiment Analysis and the Semantic Web Co-located with 14th ESWC 2017, Portroz, Slovenia, May 28, 2017. Volume 1874 of CEUR Workshop Proceedings., CEUR-WS.org (2017)
28. Riloff, E., Patwardhan, S., Wiebe, J.: Feature subsumption for opinion analysis. In: EMNLP, pp. 440–448 (2006)
29. Wilson, T., Wiebe, J., Hwa, R.: Recognizing strong and weak opinion clauses. Comput. Intell. **22**(2), 73–99 (2006)
30. Palmero Aprosio, A., Corcoglioniti, F., Dragoni, M., Rospocher, M.: Supervised opinion frames detection with RAID. In: Gandon, F., Cabrio, E., Stankovic, M., Zimmermann, A. (eds.) SemWebEval 2015. CCIS, vol. 548, pp. 251–263. Springer, Cham (2015). https://doi.org/10.1007/978-3-319-25518-7_22
31. Hatzivassiloglou, V., Wiebe, J.: Effects of adjective orientation and gradability on sentence subjectivity. In: COLING, pp. 299–305 (2000)
32. Kim, S.M., Hovy, E.H.: Crystal: analyzing predictive opinions on the web. In: EMNLP-CoNLL, pp. 1056–1064 (2007)
33. Rexha, A., Kröll, M., Dragoni, M., Kern, R.: Polarity classification for target phrases in tweets: a Word2Vec approach. In: Sack, H., Rizzo, G., Steinmetz, N., Mladenić, D., Auer, S., Lange, C. (eds.) ESWC 2016. LNCS, vol. 9989, pp. 217–223. Springer, Cham (2016). https://doi.org/10.1007/978-3-319-47602-5_40
34. Rexha, A., Kröll, M., Kern, R., Dragoni, M.: An embedding approach for microblog polarity classification. In: Dragoni, M., Recupero, D.R. (eds.) Proceedings of the 3rd International Workshop on Emotions, Modality, Sentiment Analysis and the Semantic Web co-located with 14th ESWC 2017, Portroz, Slovenia, May 28, 2017. Volume 1874 of CEUR Workshop Proceedings., CEUR-WS.org (2017)
35. Recupero, D.R., Dragoni, M., Presutti, V.: ESWC 15 challenge on concept-level sentiment analysis. In: Gandon, F., Cabrio, E., Stankovic, M., Zimmermann, A. (eds.) SemWebEval 2015. CCIS, vol. 548, pp. 211–222. Springer, Cham (2015). https://doi.org/10.1007/978-3-319-25518-7_18
36. Dragoni, M., Reforgiato Recupero, D.: Challenge on fine-grained sentiment analysis within ESWC2016. In: Sack, H., Dietze, S., Tordai, A., Lange, C. (eds.) SemWebEval 2016. CCIS, vol. 641, pp. 79–94. Springer, Cham (2016). https://doi.org/10.1007/978-3-319-46565-4_6
37. Dragoni, M., Solanki, M., Blomqvist, E., eds.: Semantic Web Challenges - 4th SemWebEval Challenge at ESWC 2017, Portoroz, Slovenia, May 28 - June 1, 2017, Revised Selected Papers. Volume 769 of Communications in Computer and Information Science. Springer (2017)
38. Jakob, N., Gurevych, I.: Extracting opinion targets in a single and cross-domain setting with conditional random fields. In: EMNLP, pp. 1035–1045 (2010)
39. Jin, W., Ho, H.H., Srihari, R.K.: Opinionminer: a novel machine learning system for web opinion mining and extraction. In: KDD, pp. 1195–1204 (2009)
40. Liu, B., Hu, M., Cheng, J.: Opinion observer: analyzing and comparing opinions on the web. In: WWW, pp. 342–351 (2005)

41. Wu, Y., Zhang, Q., Huang, X., Wu, L.: Phrase dependency parsing for opinion mining. In: EMNLP, pp. 1533–1541 (2009)
42. Su, Q., et al.: Hidden sentiment association in Chinese web opinion mining. In: WWW, pp. 959–968 (2008)
43. Dragoni, M.: NEUROSENT-PDI at semeval-2018 task 1: leveraging a multi-domain sentiment model for inferring polarity in micro-blog text. In: Apidianaki, M., Mohammad, S.M., May, J., Shutova, E., Bethard, S., Carpuat, M. (eds.) Proceedings of The 12th International Workshop on Semantic Evaluation, SemEval@NAACL-HLT, New Orleans, Louisiana, June 5–6, 2018, Association for Computational Linguistics, pp. 102–108 (2018)
44. Dragoni, M.: NEUROSENT-PDI at semeval-2018 task 3: understanding irony in social networks through a multi-domain sentiment model. In: Apidianaki, M., Mohammad, S.M., May, J., Shutova, E., Bethard, S., Carpuat, M. (eds.) Proceedings of The 12th International Workshop on Semantic Evaluation, SemEval@NAACL-HLT, New Orleans, Louisiana, June 5–6, 2018, Association for Computational Linguistics, pp. 512–519 (2018)
45. Dragoni, M., Azzini, A., Tettamanzi, A.G.B.: A novel similarity-based crossover for artificial neural network evolution. In: Schaefer, R., Cotta, C., Kołodziej, J., Rudolph, G. (eds.) PPSN 2010. LNCS, vol. 6238, pp. 344–353. Springer, Heidelberg (2010). https://doi.org/10.1007/978-3-642-15844-5_35
46. Qiu, G., Liu, B., Bu, J., Chen, C.: Opinion word expansion and target extraction through double propagation. Comput. Linguist. **37**(1), 9–27 (2011)
47. Dragoni, M., da Costa Pereira, C., Tettamanzi, A.G.B., Villata, S.: Combining argumentation and aspect-based opinion mining: the smack system. AI Commun. **31**(1), 75–95 (2018)
48. Dragoni, M.: A three-phase approach for exploiting opinion mining in computational advertising. IEEE Intell. Syst. **32**(3), 21–27 (2017)
49. Dragoni, M., Petrucci, G.: A neural word embeddings approach for multi-domain sentiment analysis. IEEE Trans. Affect. Comput. **8**(4), 457–470 (2017)
50. Dragoni, M.: Computational advertising in social networks: an opinion mining-based approach. In: Haddad, H.M., Wainwright, R.L., Chbeir, R. (eds.) Proceedings of the 33rd Annual ACM Symposium on Applied Computing, SAC 2018, Pau, France, April 09–13, 2018, ACM, pp. 1798–1804 (2018)
51. Barbosa, L., Feng, J.: Robust sentiment detection on twitter from biased and noisy data. In: COLING (Posters), pp. 36–44 (2010)
52. Bermingham, A., Smeaton, A.F.: Classifying sentiment in microblogs: is brevity an advantage? In: CIKM, pp. 1833–1836 (2010)
53. Go, A., Bhayani, R., Huang, L.: Twitter sentiment classification using distant supervision. CS224N Project Report, Standford University (2009)
54. Cambria, E., Hussain, A.: Sentic computing: a common-sense-based framework for concept-level sentiment analysis (2015)
55. Cambria, E., Hussain, A.: Sentic album: content-, concept-, and context-based online personal photo management system. Cogn. Comput. **4**(4), 477–496 (2012)
56. Wang, Q.F., Cambria, E., Liu, C.L., Hussain, A.: Common sense knowledge for handwritten chinese recognition. Cogn. Comput. **5**(2), 234–242 (2013)
57. Yoshida, Y., Hirao, T., Iwata, T., Nagata, M., Matsumoto, Y.: Transfer learning for multiple-domain sentiment analysis–identifying domain dependent/independent word polarity. AAA I, 1286–1291 (2011)
58. Ponomareva, N., Thelwall, M.: Semi-supervised vs. cross-domain graphs for sentiment analysis. In: RANLP, pp. 571–578 (2013)

59. Huang, S., Niu, Z., Shi, C.: Automatic construction of domain-specific sentiment lexicon based on constrained label propagation. Knowl.-Based Syst. **56**, 191–200 (2014)

60. Dragoni, M., da Costa Pereira, C., Tettamanzi, A.G.B., Villata, S.: Smack: an argumentation framework for opinion mining. In Kambhampati, S. (ed.) Proceedings of the Twenty-Fifth International Joint Conference on Artificial Intelligence, IJCAI 2016, New York, NY, USA, 9–15 July 2016, IJCAI/AAAI Press, pp. 4242–4243 (2016)

61. Cambria, E., Olsher, D., Rajagopal, D.: SenticNet 3: a common and common-sense knowledge base for cognition-driven sentiment analysis. In: AAAI, pp. 1515–1521 (2014)

62. Stone, P.J., Dunphy, D., Marshall, S.: The General Inquirer: A Computer Approach to Content Analysis. M.I.T. Press, Oxford, England (1966)

63. Zadeh, L.A.: Fuzzy sets. Inf. Control **8**, 338–353 (1965)

64. Manning, C.D., Surdeanu, M., Bauer, J., Finkel, J., Bethard, S.J., McClosky, D.: The Stanford CoreNLP natural language processing toolkit. In: Proceedings of 52nd Annual Meeting of the Association for Computational Linguistics: System Demonstrations, Baltimore, Maryland, Association for Computational Linguistics, pp. 55–60 (2014)

65. van Rijsbergen, C.J.: Information Retrieval. Butterworth, London (1979)

66. Zadeh, L.A.: The concept of a linguistic variable and its application to approximate reasoning - I. Inf. Sci. **8**(3), 199–249 (1975)

67. Hellendoorn, H., Thomas, C.: Defuzzification in fuzzy controllers. Intell. Fuzzy Syst. **1**, 109–123 (1993)

68. Dragoni, M., Tettamanzi, A., da Costa Pereira, C.: Dranziera: an evaluation protocol for multi-domain opinion mining. In: Chair, N.C.C., Choukri, K., Declerck, T., Goggi, S., Grobelnik, M., Maegaard, B., Mariani, J., Mazo, H., Moreno, A., Odijk, J., Piperidis, S. (eds.) Proceedings of the Tenth International Conference on Language Resources and Evaluation (LREC 2016), Paris, France, European Language Resources Association (ELRA) (2016)

The CLAUSY System at ESWC-2018 Challenge on Semantic Sentiment Analysis

Andi Rexha[1], Mark Kröll[1], Mauro Dragoni[2(✉)], and Roman Kern[1]

[1] Know-Center GmbH Graz, Graz, Austria
{arexha,mkroell,rkern}@know-center.at
[2] FBK-IRST, Trento, Italy
dragoni@fbk.eu

Abstract. With different social media and commercial platforms, users express their opinion about products in a textual form. Automatically extracting the polarity(i.e. whether the opinion is positive or negative) of a user can be useful for both actors: the online platform incorporating the feedback to improve their product as well as the client who might get recommendations according to his or her preferences. Different approaches for tackling the problem, have been suggested mainly using syntactic features. The "Challenge on Semantic Sentiment Analysis" aims to go beyond the word-level analysis by using semantic information. In this paper we propose a novel approach by employing the semantic information of grammatical unit called preposition. We try to derive the *target* of the review from the *summary information*, which serves as an input to identify the proposition in it. Our implementation relies on the hypothesis that the proposition expressing the target of the summary, usually containing the main polarity information.

1 Introduction

User's opinions can be found in various social media platforms and online stores in textual form. The length and style of the text can vary substantially, ranging from short twitter messages to longer book reviews. They also refer to different aspects, from politics, to pictures and comercial products. The nature of these opinions change the way to analyze the text from automatic polarity detection systems. Twitter messages aren't expressed using syntactically correct text and require a different preprocessing than, for example, book reviews. For the "Challenge on Semantic Sentiment Analysis" the task (with the winners of recent years [1,7,24,56]) is to detect the polarity of user's opinions of products on Amazon.com reviews. The dataset [23] consists of a set of summaries about different topics. The review is compound by an *id*, a *summary* and a *textual description* and is represented in a XML format as shown in the Example 1.

Since the summary text is expressed in well formed text, Natural Language Processing (NLP) tools can be used to preprocess and analyze those reviews.

© Springer Nature Switzerland AG 2018
D. Buscaldi et al. (Eds.): SemWebEval 2018, CCIS 927, pp. 186–196, 2018.
https://doi.org/10.1007/978-3-030-00072-1_15

```
<summary>Transformers</summary>
<text>By most accounts, the Michael Bay-directed Transformers
    films to date films to date are not very good, but that
    hasnt stopped them from making gobs and gobs of cash.
</text>
<polarity>positive</polarity>
```

Example 1: Example of a single entry in the dataset provided by the challenge

For this challenge we use a two step approach. In the first step we isolate the syntactic information(proposition) in which the summary is expressed. In the second step we use a supervised approach in order to classify the reviews in positive or negative polarity.

The paper is organized in four sections. In Sect. 3 we detail the approach used for the two steps and the features used for the supervised task. Finally, Sect. 4 describes the partial results and the discussion the advantages and drawback of the approach.

2   Related Work

The topic of sentiment analysis has been studied extensively in the literature [38], where several techniques have been proposed and validated.

Machine learning techniques are the most common approaches used for addressing this problem, given that any existing supervised methods can be applied to sentiment classification. For instance, in [43], the authors compared the performance of Naive-Bayes, Maximum Entropy, and Support Vector Machines in sentiment analysis on different features like considering only unigrams, bigrams, combination of both, incorporating parts of speech and position information or by taking only adjectives. Moreover, beside the use of standard machine learning method, researchers have also proposed several custom techniques specifically for sentiment classification, like the use of adapted score function based on the evaluation of positive or negative words in product reviews [9], as well as by defining weighting schemata for enhancing classification accuracy [41].

An obstacle to research in this direction is the need of labeled training data, whose preparation is a time-consuming activity. Therefore, in order to reduce the labeling effort, opinion words have been used for training procedures. In [49,60], the authors used opinion words to label portions of informative examples for training the classifiers. Opinion words have been exploited also for improving the accuracy of sentiment classification, as presented in [40], where a framework incorporating lexical knowledge in supervised learning to enhance accuracy has been proposed. Opinion words have been used also for unsupervised learning approaches like the one presented in [59].

Another research direction concerns the exploitation of discourse-analysis techniques. [57] discusses some discourse-based supervised and unsupervised approaches for opinion analysis; while in [61], the authors present an approach to identify discourse relations.

The approaches presented above are applied at the document-level [11,27, 29,44,51,53], i.e., the polarity value is assigned to the entire document content. However, in some case, for improving the accuracy of the sentiment classification, a more fine-grained analysis of a document is needed. Hence, the sentiment classification of the single sentences, has to be performed. In the literature, we may find approaches ranging from the use of fuzzy logic [20,24,25,45,46] to the use of aggregation techniques [8] for computing the score aggregation of opinion words. In the case of sentence-level sentiment classification, two different sub-tasks have to be addressed: (i) to determine if the sentence is subjective or objective, and (ii) in the case that the sentence is subjective, to determine if the opinion expressed in the sentence is positive, negative, or neutral. The task of classifying a sentence as subjective or objective, called "subjectivity classification", has been widely discussed in the literature [28,30,55,63] and systems implementing the capabilities of identifying opinion's holder, target, and polarity have been presented [1]. Once subjective sentences are identified, the same methods as for sentiment classification may be applied. For example, in [32] the authors consider gradable adjectives for sentiment spotting; while in [36,52,54] the authors built models to identify some specific types of opinions.

In the last years, with the growth of product reviews, the use of sentiment analysis techniques was the perfect floor for validating them in marketing activities [21,22,50]. However, the issue of improving the ability of detecting the different opinions concerning the same product expressed in the same review became a challenging problem. Such a task has been faced by introducing "aspect" extraction approaches that were able to extract, from each sentence, which is the aspect the opinion refers to. In the literature, many approaches have been proposed: conditional random fields (CRF) [34], hidden Markov models (HMM) [35], sequential rule mining [37], dependency tree kernels [64], clustering [58], neural networks [14,15], and genetic algorithms [16]. In [18,48], two methods were proposed to extract both opinion words and aspects simultaneously by exploiting some syntactic relations of opinion words and aspects.

A particular attention should be given also to the application of sentiment analysis in social networks [12,13,19]. More and more often, people use social networks for expressing their moods concerning their last purchase or, in general, about new products. Such a social network environment opened up new challenges due to the different ways people express their opinions, as described by [2,3], who mention "noisy data" as one of the biggest hurdles in analyzing social network texts.

One of the first studies on sentiment analysis on micro-blogging websites has been discussed in [31], where the authors present a distant supervision-based approach for sentiment classification.

At the same time, the social dimension of the Web opens up the opportunity to combine computer science and social sciences to better recognize, interpret, and process opinions and sentiments expressed over it. Such multi-disciplinary approach has been called *sentic computing* [6]. Application domains where sentic computing has already shown its potential are the cognitive-inspired classification of images [5], of texts in natural language, and of handwritten text [62].

Finally, an interesting recent research direction is domain adaptation, as it has been shown that sentiment classification is highly sensitive to the domain from which the training data is extracted. A classifier trained using opinionated documents from one domain often performs poorly when it is applied or tested on opinionated documents from another domain, as we demonstrated through the example presented in Sect. 1. The reason is that words and even language constructs used in different domains for expressing opinions can be quite different. To make matters worse, the same word in one domain may have positive connotations, but in another domain may have negative ones; therefore, domain adaptation is needed. In the literature, different approaches related to the Multi-Domain sentiment analysis have been proposed. Briefly, two main categories may be identified: (i) the transfer of learned classifiers across different domains [4,42,65], and (ii) the use of propagation of labels through graph structures [17,25,33,47].

All approaches presented above are based on the use of statistical techniques for building sentiment models. The exploitation of semantic information is not taken into account. In this work, we proposed a first version of a semantic-based approach preserving the semantic relationships between the terms of each sentence in order to exploit them either for building the model and for estimating document polarity. The proposed approach, falling into the multi-domain sentiment analysis category, instead of using pre-determined polarity information associated with terms, it learns them directly from domain-specific documents. Such documents are used for training the models used by the system.

3   Approach and Features

Each review is composed of a *summary* and a *textual information*. One or more sentences form the textual information of the summary contain the detailed specification of the user experience. We base our approach in the hypothesis that the summary is extended in the textual information and its "isolated" content contains the main polarity information. After preprocessing the textual summary with NLP tools, we annotate the words of the summary in each sentence. Later we extract the most "prominent" sentence (to be defined in Sect. 3.2) which contains the main target of the summary. From the "prominent" sentence we select the "best fitting" proposition (we define it in Sect. 3.3) which contains the summary. From the proposition, we extract the polarity of each word and encode the distribution of the polarity in the proposition as features. As a final step we train a classifier in order to predict the polarity of the whole tweet. Recapping, the approach can be split in the following steps:

- Preprocessing
- Extract the prominent sentence
- Extract the prominent proposition
- Polarity extraction and feature encoding

Below, we describe each of these steps in more detail. For illustration purposes we get the following example:

> **Summary:** Typical movie of Al Pacino
> **Text:** This was a very good movie from Al Pacino but the music wasn't that nice. Just think about how bad other movies are! The music doesn't play any role for my review!

Example 2: Example of a summary and review

3.1 Preprocessing

For preprocessing the reviews, we select the Stanford Core NLP tool [39]. For each review we annotate all sentences, words and parse the syntactic dependency graph. As a final step we annotate the text in the review with the tokens from the summary.

In the example 2 this would be: This was a very good `movie` from `Al Pacino` but the music wasn't that nice.

3.2 Extract the Prominent Sentence

From the annotated text of the review we need to select the sentence best matching with the summary. We define the most "prominent" one as the sentence which contains more terms in a TermFrequency-InverseSentenceFrequency of the term. So, for each term of the summary we calculate it's frequency (i.e. the number of times it occurs in each sentence) and it's inverse sentence frequency (i.e. the inverse fraction of documents containing the word). This formula reflects the tf-idf(term frequency-inverse document frequency) score, but applied to the sentences, and we consider it a tf-isf. In the former example, the first sentence would be selected due to it containing two annotated words.

3.3 Extract the Prominent Proposition

For the "best fit" sentence we try to extract the *proposition* which capture best the main information. As a first step, we extract the propositions composing the sentence. We use the well known Open Information Extraction tool, ClausIE [10]. It extracts relation of the form (subject, predicate, object) called propositions.

Returning to the example of the prominent sentence "This was a very good `movie` from `Al Pacino` but the music wasn't that nice.", it can be splitted in the following propositions:

– This was a very good movie
– This was a very good movie from Al Pacino
– the music wasn't that nice.

We define the "best matching" proposition as the shortest one (in terms of words) containing most of the terms in the summary. This mean that the selected one in our example would be: This was a very good movie from Al Pacino .

3.4   Features

In this challenge we use polarity features extracted from SentiWordNet [26]. SentiWordNet is a thesaurus which contains polarity information about words. To each word it is assigned a score between -1 and 1, which indicates whether the word has a negative or positive polarity. We model the proposition as a function of the sentiment expressed in the words. More precisely we identify the polarity of each word in the "best fit" proposition. We express the features of the "best fit" proposition as maximum, minimum, arithmetic mean, and standard deviation of the polarities of the words. As a additional feature we use the number of negation words expressed in the "best matching" sentence.

4   Results and Discussion

We try to learn our model from the features we have extracted and predict new unseen reviews. We use a Logistic Regression to learn from the results. In the Table 1 we present the precision, recall and F1-measure of the 10 fold cross-validation. While Tables 2 and 3 shows the full results of Tasks #1 and #2 participants.

Table 1: Results from a 10 fold cross validation in the training dataset.

	Precision	Recall	F1-measure
Positive	0.629	0.823	0.713
Negative	0.744	0.515	0.608
Average	0.686	0.669	0.661

As we can see from the tables, the results from the cross validation are not as good as expected. After an analyses of the dataset we believe that this discrepancy of the results from the expectation might be caused by the false assumption that the summary of the review is also expressed in the text.

Table 2: Results obtained by the participant to Task #1.

System	F-Measure
Guangyuan et. al	0.9643
Atzori et. al	0.9561
Federici et. al	0.9356
Dragoni et. al	0.9228
Dragoni et. al	0.8823
Rexha et. al	**0.8743**
Indurthi et. al	0.8203
Petrucci et. al	0.7153
Chaudhuri et. al	0.5243

Table 3: Results obtained by the participant to Task #2.

System	F-Measure
Federici et. al	0.5682
Dragoni et. al	0.5244
Rexha et. al	**0.4598**

Acknowledgment. This work is funded by the KIRAS program of the Austrian Research Promotion Agency (FFG) (project number 840824). The Know-Center is funded within the Austrian COMET Program under the auspices of the Austrian Ministry of Transport, Innovation and Technology, the Austrian Ministry of Economics and Labour and by the State of Styria. COMET is managed by the Austrian Research Promotion Agency FFG

References

1. Palmero Aprosio, A., Corcoglioniti, F., Dragoni, M., Rospocher, M.: Supervised opinion frames detection with RAID. In: Gandon, F., Cabrio, E., Stankovic, M., Zimmermann, A. (eds.) SemWebEval 2015. CCIS, vol. 548, pp. 251–263. Springer, Cham (2015). https://doi.org/10.1007/978-3-319-25518-7_22
2. Barbosa, L., Feng, J.: Robust sentiment detection on twitter from biased and noisy data. In: COLING (Posters), pp. 36–44 (2010)
3. Bermingham, A., Smeaton, A.F.: Classifying sentiment in microblogs: is brevity an advantage? In: CIKM, pp. 1833–1836 (2010)
4. Blitzer, J., Dredze, M., Pereira, F.: Biographies, bollywood, boom-boxes and blenders: Domain adaptation for sentiment classification. In: ACL, pp. 187–205 (2007)
5. Cambria, E., Hussain, A.: Sentic album: Content-, concept-, and context-based online personal photo management system. Cogn. Comput. **4**(4), 477–496 (2012)
6. Cambria, E., Hussain, A.: Sentic computing: a common-sense-based framework for concept-level sentiment analysis (2015)

7. Chung, J.K.-C., Wu, C.-E., Tsai, R.T.-H.: Polarity detection of online reviews using sentiment concepts: NCU IISR team at ESWC-14 challenge on concept-level sentiment analysis. In: Presutti, V., Stankovic, M., Cambria, E., Cantador, I., Di Iorio, A., Di Noia, T., Lange, C., Reforgiato Recupero, D., Tordai, A. (eds.) SemWebEval 2014. CCIS, vol. 475, pp. 53–58. Springer, Cham (2014). https://doi.org/10.1007/978-3-319-12024-9_7

8. da Costa Pereira, C., Dragoni, M., Pasi, G.: A prioritized "and" aggregation operator for multidimensional relevance assessment. In: Serra, R., Cucchiara, R. (eds.) AI*IA 2009. LNCS (LNAI), vol. 5883, pp. 72–81. Springer, Heidelberg (2009). https://doi.org/10.1007/978-3-642-10291-2_8

9. Dave, K., Lawrence, S., Pennock, D.M.: Mining the peanut gallery: opinion extraction and semantic classification of product reviews. In: WWW, pp. 519–528 (2003)

10. Del Corro, L., Gemulla, R.: Clausie: clause-based open information extraction. In: Proceedings of the 22Nd International Conference on World Wide Web, pp. 355–366. WWW '13, ACM, New York, NY, USA (2013). https://doi.org/10.1145/2488388.2488420

11. Dragoni, M.: Shellfbk: an information retrieval-based system for multi-domain sentiment analysis. In: Proceedings of the 9th International Workshop on Semantic Evaluation, pp. 502–509. SemEval '2015, Association for Computational Linguistics, Denver, Colorado (June 2015)

12. Dragoni, M.: A three-phase approach for exploiting opinion mining in computational advertising. IEEE Intell. Syst. **32**(3), 21–27 (2017). https://doi.org/10.1109/MIS.2017.46

13. Dragoni, M.: Computational advertising in social networks: an opinion mining-based approach. In: Haddad, H.M., Wainwright, R.L., Chbeir, R. (eds.) Proceedings of the 33rd Annual ACM Symposium on Applied Computing, SAC 2018, Pau, France, April 09–13, 2018, pp. 1798–1804. ACM (2018), https://doi.org/10.1145/3167132.3167324

14. Dragoni, M.: NEUROSENT-PDI at semeval-2018 task 1: Leveraging a multi-domain sentiment model for inferring polarity in micro-blog text. In: Apidianaki, M., Mohammad, S.M., May, J., Shutova, E., Bethard, S., Carpuat, M. (eds.) Proceedings of The 12th International Workshop on Semantic Evaluation, SemEval@NAACL-HLT, New Orleans, Louisiana, June 5–6, 2018, pp. 102–108. Association for Computational Linguistics (2018). https://aclanthology.info/papers/S18-1013/s18-1013

15. Dragoni, M.: NEUROSENT-PDI at semeval-2018 task 3: understanding irony in social networks through a multi-domain sentiment model. In: Apidianaki, M., Mohammad, S.M., May, J., Shutova, E., Bethard, S., Carpuat, M. (eds.) Proceedings of The 12th International Workshop on Semantic Evaluation, SemEval@NAACL-HLT, New Orleans, Louisiana, June 5–6, 2018, pp. 512–519. Association for Computational Linguistics (2018). https://aclanthology.info/papers/S18-1083/s18-1083

16. Dragoni, M., Azzini, A., Tettamanzi, A.G.B.: A novel similarity-based crossover for artificial neural network evolution. In: Schaefer, R., Cotta, C., Kołodziej, J., Rudolph, G. (eds.) PPSN 2010. LNCS, vol. 6238, pp. 344–353. Springer, Heidelberg (2010). https://doi.org/10.1007/978-3-642-15844-5_35

17. Dragoni, M., da Costa Pereira, C., Tettamanzi, A.G.B., Villata, S.: Smack: an argumentation framework for opinion mining. In: Kambhampati, S. (ed.) Proceedings of the Twenty-Fifth International Joint Conference on Artificial Intelligence,

IJCAI 2016, New York, NY, USA, 9–15 July 2016, pp. 4242–4243. IJCAI/AAAI Press (2016). http://www.ijcai.org/Abstract/16/641

18. Dragoni, M., da Costa Pereira, C., Tettamanzi, A.G.B., Villata, S.: Combining argumentation and aspect-based opinion mining: the smack system. AI Commun. **31**(1), 75–95 (2018). https://doi.org/10.3233/AIC-180752

19. Dragoni, M., Petrucci, G.: A neural word embeddings approach for multi-domain sentiment analysis. IEEE Trans. Affect. Comput. **8**(4), 457–470 (2017). https://doi.org/10.1109/TAFFC.2017.2717879

20. Dragoni, M., Petrucci, G.: A fuzzy-based strategy for multi-domain sentiment analysis. Int. J. Approx. Reason. **93**, 59–73 (2018). https://doi.org/10.1016/j.ijar.2017.10.021

21. Dragoni, M., Reforgiato Recupero, D.: Challenge on fine-grained sentiment analysis within ESWC2016. In: Sack, H., Dietze, S., Tordai, A., Lange, C. (eds.) SemWebEval 2016. CCIS, vol. 641, pp. 79–94. Springer, Cham (2016). https://doi.org/10.1007/978-3-319-46565-4_6

22. Dragoni, M., Solanki, M., Blomqvist, E. (eds.): SemWebEval 2017. CCIS, vol. 769. Springer, Cham (2017). https://doi.org/10.1007/978-3-319-69146-6

23. Dragoni, M., Tettamanzi, A., da Costa Pereira, C.: Dranziera: an evaluation protocol for multi-domain opinion mining. In: Chair, N.C.C., et al. (eds.) Proceedings of the Tenth International Conference on Language Resources and Evaluation (LREC 2016). European Language Resources Association (ELRA), Paris, France (may 2016)

24. Dragoni, M., Tettamanzi, A.G.B., da Costa Pereira, C.: A fuzzy system for concept-level sentiment analysis. In: Presutti, V., Stankovic, M., Cambria, E., Cantador, I., Di Iorio, A., Di Noia, T., Lange, C., Reforgiato Recupero, D., Tordai, A. (eds.) SemWebEval 2014. CCIS, vol. 475, pp. 21–27. Springer, Cham (2014). https://doi.org/10.1007/978-3-319-12024-9_2

25. Dragoni, M., Tettamanzi, A.G., da Costa Pereira, C.: Propagating and aggregating fuzzy polarities for concept-level sentiment analysis. Cogn. Comput. **7**(2), 186–197 (2015). https://doi.org/10.1007/s12559-014-9308-6

26. Esuli, A., Sebastiani, F.: Sentiwordnet: a publicly available lexical resource for opinion mining. In: Proceedings of the 5th Conference on Language Resources and Evaluation, pp. 417–422. LREC06 (2006)

27. Federici, M., Dragoni, M.: A knowledge-based approach for aspect-based opinion mining. In: Sack, H., Dietze, S., Tordai, A., Lange, C. (eds.) SemWebEval 2016. CCIS, vol. 641, pp. 141–152. Springer, Cham (2016). https://doi.org/10.1007/978-3-319-46565-4_11

28. Federici, M., Dragoni, M.: Towards unsupervised approaches for aspects extraction. In: Dragoni, M., Recupero, D.R., Denecke, K., Deng, Y., Declerck, T. (eds.) Joint Proceedings of the 2th Workshop on Emotions, Modality, Sentiment Analysis and the Semantic Web and the 1st International Workshop on Extraction and Processing of Rich Semantics from Medical Texts co-located with ESWC 2016, Heraklion, Greece, May 29, 2016. CEUR Workshop Proceedings, vol. 1613. CEUR-WS.org (2016). http://ceur-ws.org/Vol-1613/paper_2.pdf

29. Federici, M., Dragoni, M.: Aspect-based opinion mining using knowledge bases. In: Dragoni, M., Solanki, M., Blomqvist, E. (eds.) SemWebEval 2017. CCIS, vol. 769, pp. 133–147. Springer, Cham (2017). https://doi.org/10.1007/978-3-319-69146-6_13

30. Federici, M., Dragoni, M.: A branching strategy for unsupervised aspect-based sentiment analysis. In: Dragoni, M., Recupero, D.R. (eds.) Proceedings of the 3rd

International Workshop at ESWC on Emotions, Modality, Sentiment Analysis and the Semantic Web co-located with 14th ESWC 2017, Portroz, Slovenia, May 28, 2017. CEUR Workshop Proceedings, vol. 1874. CEUR-WS.org (2017). http://ceur-ws.org/Vol-1874/paper_6.pdf

31. Go, A., Bhayani, R., Huang, L.: Twitter sentiment classification using distant supervision. CS224N Project Report, Standford University (2009)

32. Hatzivassiloglou, V., Wiebe, J.: Effects of adjective orientation and gradability on sentence subjectivity. In: COLING, pp. 299–305 (2000)

33. Huang, S., Niu, Z., Shi, C.: Automatic construction of domain-specific sentiment lexicon based on constrained label propagation. Knowl.-Based Syst. **56**, 191–200 (2014)

34. Jakob, N., Gurevych, I.: Extracting opinion targets in a single and cross-domain setting with conditional random fields. In: EMNLP, pp. 1035–1045 (2010)

35. Jin, W., Ho, H.H., Srihari, R.K.: Opinionminer: a novel machine learning system for web opinion mining and extraction. In: KDD. pp, 1195–1204 (2009)

36. Kim, S.M., Hovy, E.H.: Crystal: analyzing predictive opinions on the web. In: EMNLP-CoNLL, pp. 1056–1064 (2007)

37. Liu, B., Hu, M., Cheng, J.: Opinion observer: analyzing and comparing opinions on the web. In: WWW, pp. 342–351 (2005)

38. Liu, B., Zhang, L.: A survey of opinion mining and sentiment analysis. In: Aggarwal, C.C., Zhai, C.X. (eds.) Mining Text Data, pp. 415–463. Springer, Berlin (2012)

39. Manning, C.D., Surdeanu, M., Bauer, J., Finkel, J. Inc, P., Bethard, S.J., Mcclosky, D.: The stanford corenlp natural language processing toolkit. In. In Proceedings of the 52nd Annual Meeting of the Association for Computational Linguistics: System Demonstrations, pp. 55–60 (2014)

40. Melville, P., Gryc, W., Lawrence, R.D.: Sentiment analysis of blogs by combining lexical knowledge with text classification. In: KDD, pp. 1275–1284 (2009)

41. Paltoglou, G., Thelwall, M.: A study of information retrieval weighting schemes for sentiment analysis. In: ACL, pp. 1386–1395 (2010)

42. Pan, S.J., Ni, X., Sun, J.T., Yang, Q., Chen, Z.: Cross-domain sentiment classification via spectral feature alignment. In: WWW, pp. 751–760 (2010)

43. Pang, B., Lee, L.: A sentimental education: Sentiment analysis using subjectivity summarization based on minimum cuts. In: ACL, pp. 271–278 (2004)

44. Petrucci, G., Dragoni, M.: An information retrieval-based system for multi-domain sentiment analysis. In: Gandon, F., Cabrio, E., Stankovic, M., Zimmermann, A. (eds.) SemWebEval 2015. CCIS, vol. 548, pp. 234–243. Springer, Cham (2015). https://doi.org/10.1007/978-3-319-25518-7_20

45. Petrucci, G., Dragoni, M.: The IRMUDOSA system at ESWC-2016 challenge on semantic sentiment analysis. In: Sack, H., Dietze, S., Tordai, A., Lange, C. (eds.) SemWebEval 2016. CCIS, vol. 641, pp. 126–140. Springer, Cham (2016). https://doi.org/10.1007/978-3-319-46565-4_10

46. Petrucci, G., Dragoni, M.: The IRMUDOSA system at ESWC-2017 challenge on semantic sentiment analysis. In: Dragoni, M., Solanki, M., Blomqvist, E. (eds.) SemWebEval 2017. CCIS, vol. 769, pp. 148–165. Springer, Cham (2017). https://doi.org/10.1007/978-3-319-69146-6_14

47. Ponomareva, N., Thelwall, M.: Semi-supervised vs. cross-domain graphs for sentiment analysis. In: RANLP, pp. 571–578 (2013)

48. Qiu, G., Liu, B., Bu, J., Chen, C.: Opinion word expansion and target extraction through double propagation. Comput. Linguist. **37**(1), 9–27 (2011)

49. Qiu, L., Zhang, W., Hu, C., Zhao, K.: Selc: a self-supervised model for sentiment classification. In: CIKM, pp. 929–936 (2009)

50. Recupero, D.R., Dragoni, M., Presutti, V.: ESWC 15 challenge on concept-level sentiment analysis. In: Gandon, F., Cabrio, E., Stankovic, M., Zimmermann, A. (eds.) SemWebEval 2015. CCIS, vol. 548, pp. 211–222. Springer, Cham (2015). https://doi.org/10.1007/978-3-319-25518-7_18

51. Rexha, A., Kröll, M., Dragoni, M., Kern, R.: Exploiting propositions for opinion mining. In: Sack, H., Dietze, S., Tordai, A., Lange, C. (eds.) SemWebEval 2016. CCIS, vol. 641, pp. 121–125. Springer, Cham (2016). https://doi.org/10.1007/978-3-319-46565-4_9

52. Rexha, A., Kröll, M., Dragoni, M., Kern, R.: Polarity classification for target phrases in tweets: a Word2Vec approach. In: Sack, H., Rizzo, G., Steinmetz, N., Mladenić, D., Auer, S., Lange, C. (eds.) ESWC 2016. LNCS, vol. 9989, pp. 217–223. Springer, Cham (2016). https://doi.org/10.1007/978-3-319-47602-5_40

53. Rexha, A., Kröll, M., Dragoni, M., Kern, R.: Opinion mining with a clause-based approach. In: Dragoni, M., Solanki, M., Blomqvist, E. (eds.) SemWebEval 2017. CCIS, vol. 769, pp. 166–175. Springer, Cham (2017). https://doi.org/10.1007/978-3-319-69146-6_15

54. Rexha, A., Kröll, M., Kern, R., Dragoni, M.: An embedding approach for microblog polarity classification. In: Dragoni, M., Recupero, D.R. (eds.) Proceedings of the 3rd International Workshop on Emotions, Modality, Sentiment Analysis and the Semantic Web co-located with 14th ESWC 2017, Portroz, Slovenia, May 28, 2017. CEUR Workshop Proceedings, vol. 1874 (2017). www.CEUR-WS.org

55. Riloff, E., Patwardhan, S., Wiebe, J.: Feature subsumption for opinion analysis. In: EMNLP, pp. 440–448 (2006)

56. Schouten, K., Frasincar, F.: The benefit of concept-based features for sentiment analysis. In: Gandon, F., Cabrio, E., Stankovic, M., Zimmermann, A. (eds.) SemWebEval 2015. CCIS, vol. 548, pp. 223–233. Springer, Cham (2015). https://doi.org/10.1007/978-3-319-25518-7_19

57. Somasundaran, S.: Discourse-level relations for Opinion Analysis. Ph.D. thesis, University of Pittsburgh (2010)

58. Su, Q., Xu, X., Guo, H., Guo, Z., Wu, X., Zhang, X., Swen, B., Su, Z.: Hidden sentiment association in chinese web opinion mining. In: WWW, pp. 959–968 (2008)

59. Taboada, M., Brooke, J., Tofiloski, M., Voll, K.D., Stede, M.: Lexicon-based methods for sentiment analysis. Comput. Linguist. **37**(2), 267–307 (2011)

60. Tan, S., Wang, Y., Cheng, X.: Combining learn-based and lexicon-based techniques for sentiment detection without using labeled examples. In: SIGIR, pp. 743–744 (2008)

61. Wang, H., Zhou, G.: Topic-driven multi-document summarization. In: IALP, pp. 195–198 (2010)

62. Wang, Q.F., Cambria, E., Liu, C.L., Hussain, A.: Common sense knowledge for handwritten chinese recognition. Cogn. Comput. **5**(2), 234–242 (2013)

63. Wilson, T., Wiebe, J., Hwa, R.: Recognizing strong and weak opinion clauses. Comput. Intell. **22**(2), 73–99 (2006)

64. Wu, Y., Zhang, Q., Huang, X., Wu, L.: Phrase dependency parsing for opinion mining. In: EMNLP, pp. 1533–1541 (2009)

65. Yoshida, Y., Hirao, T., Iwata, T., Nagata, M., Matsumoto, Y.: Transfer learning for multiple-domain sentiment analysis–identifying domain dependent/independent word polarity. In: AAAI, pp. 1286–1291 (2011)

The NeuroSent System at ESWC-2018 Challenge on Semantic Sentiment Analysis

Mauro Dragoni[✉]

Fondazione Bruno Kessler, Trento, Italy
dragoni@fbk.eu

Abstract. Multi-domain sentiment analysis consists in estimating the polarity of a given text by exploiting domain-specific information. One of the main issues common to the approaches discussed in the literature is their poor capabilities of being applied on domains which are different from those used for building the opinion model. In this paper, we will present an approach exploiting the linguistic overlap between domains to build sentiment models supporting polarity inference for documents belonging to every domain. Word embeddings together with a deep learning architecture have been implemented for enabling the building of multi-domain sentiment model. The proposed technique is validated by following the Dranziera protocol in order to ease the repeatability of the experiments and the comparison of the results. The outcomes demonstrate the effectiveness of the proposed approach and also set a plausible starting point for future work.

1 Introduction

Sentiment Analysis is a natural language processing (NLP) task [1] which aims at classifying documents according to the opinion expressed about a given subject [2].

Many works available in the literature address the sentiment analysis problem without distinguishing domain specific information of documents when sentiment models are built. The necessity of investigating this problem from a multi-domain perspective is led by the different influence that a term might have in different contexts.

The idea of adapting terms polarity to different domains emerged only in the last decade [3]. Multi-domain sentiment analysis approaches discussed in the literature (surveyed in Sect. 2) focus on building models for transferring information between pairs of domains [4]. While on the one hand such approaches allow to propagate specific domain information to others, their drawback is the necessity of building new transfer models every time a new domain has to be analyzed. Thus, such approaches do not have a great generalization capability of analyzing texts, because transfer models are limited to the N domains used for building the models.

D. Buscaldi et al. (Eds.): SemWebEval 2018, CCIS 927, pp. 197–215, 2018.
https://doi.org/10.1007/978-3-030-00072-1_16

The contribution presented in this paper aims at addressing the challenge of working in a multi-domain environment. The described approach, implemented into the NeuroSent tool, is based on the following pillars:

- the use of word embeddings for representing each word contained in raw sentences;
- the word embeddings are generated from an opinion-based corpus instead of a general purpose one (like news or Wikipedia);
- the design of a deep learning technique exploiting the generated word embeddings for training the sentiment model;
- the use of multiple output layers for combining domain overlap scores with domain-specific polarity predictions.

The last point enables the exploitation of linguistic overlaps between domains, which can be considered one of the pivotal assets of our approach. This way, the overall polarity of a document is computed by aggregating, for each domain, the domain-specific polarity value multiplied by a belonging degree representing the overlap between the embedded representation of the whole document and the domain itself. The use of this strategy ease the validation of our approach from two perspectives: (i) to measure the effectiveness of our model on the domains used for creating the model itself, and (ii) to observe how our model behaves in classifying document coming from domains different from the ones adopted for building the model (i.e. generalization of our approach). This point represents an innovative aspect with respect to the state of the art of multi-domain sentiment analysis.

2   Related Work

The topic of sentiment analysis has been studied extensively in the literature [5], where several techniques have been proposed and validated.

Machine learning techniques are the most common approaches used for addressing this problem, given that any existing supervised methods can be applied to sentiment classification. For instance, in [6], the authors compared the performance of Naive-Bayes, Maximum Entropy, and Support Vector Machines in sentiment analysis on different features like considering only unigrams, bigrams, combination of both, incorporating parts of speech and position information or by taking only adjectives. Moreover, beside the use of standard machine learning method, researchers have also proposed several custom techniques specifically for sentiment classification, like the use of adapted score function based on the evaluation of positive or negative words in product reviews [7], as well as by defining weighting schemata for enhancing classification accuracy [8].

An obstacle to research in this direction is the need of labeled training data, whose preparation is a time-consuming activity. Therefore, in order to reduce the labeling effort, opinion words have been used for training procedures. In [9,10], the authors used opinion words to label portions of informative examples for training the classifiers. Opinion words have been exploited also for improving the accuracy of sentiment classification, as presented in [11], where a framework incorporating lexical knowledge in supervised learning to enhance accuracy has been proposed. Opinion words have been used also for unsupervised learning approaches like the one presented in [12].

Another research direction concerns the exploitation of discourse-analysis techniques. [13] discusses some discourse-based supervised and unsupervised approaches for opinion analysis; while in [14], the authors present an approach to identify discourse relations.

The approaches presented above are applied at the document-level [15–20], i.e., the polarity value is assigned to the entire document content. However, in some case, for improving the accuracy of the sentiment classification, a more fine-grained analysis of a document is needed. Hence, the sentiment classification of the single sentences, has to be performed. In the literature, we may find approaches ranging from the use of fuzzy logic [21–25] to the use of aggregation techniques [26] for computing the score aggregation of opinion words. In the case of sentence-level sentiment classification, two different sub-tasks have to be addressed: (i) to determine if the sentence is subjective or objective, and (ii) in the case that the sentence is subjective, to determine if the opinion expressed in the sentence is positive, negative, or neutral. The task of classifying a sentence as subjective or objective, called "subjectivity classification", has been widely discussed in the literature [27–30] and systems implementing the capabilities of identifying opinion's holder, target, and polarity have been presented [31]. Once subjective sentences are identified, the same methods as for sentiment classification may be applied. For example, in [32] the authors consider gradable adjectives for sentiment spotting; while in [33–35] the authors built models to identify some specific types of opinions.

In the last years, with the growth of product reviews, the use of sentiment analysis techniques was the perfect floor for validating them in marketing activities [36–38]. However, the issue of improving the ability of detecting the different opinions concerning the same product expressed in the same review became a challenging problem. Such a task has been faced by introducing "aspect" extraction approaches that were able to extract, from each sentence, which is the aspect the opinion refers to. In the literature, many approaches have been proposed: conditional random fields (CRF) [39], hidden Markov models (HMM) [40], sequential rule mining [41], dependency tree kernels [42], clustering [43], neural networks [44, 45], and genetic algorithms [46]. In [47, 48], two methods were proposed to extract both opinion words and aspects simultaneously by exploiting some syntactic relations of opinion words and aspects.

A particular attention should be given also to the application of sentiment analysis in social networks [49–51]. More and more often, people use social networks for expressing their moods concerning their last purchase or, in general, about new products. Such a social network environment opened up new challenges due to the different ways people express their opinions, as described by [52,53], who mention "noisy data" as one of the biggest hurdles in analyzing social network texts.

One of the first studies on sentiment analysis on micro-blogging websites has been discussed in [54], where the authors present a distant supervision-based approach for sentiment classification.

At the same time, the social dimension of the Web opens up the opportunity to combine computer science and social sciences to better recognize, interpret, and process opinions and sentiments expressed over it. Such multi-disciplinary approach has been called *sentic computing* [55]. Application domains where sentic computing has already shown its potential are the cognitive-inspired classification of images [56], of texts in natural language, and of handwritten text [57].

Finally, an interesting recent research direction is domain adaptation, as it has been shown that sentiment classification is highly sensitive to the domain from which the training data is extracted. A classifier trained using opinionated documents from one domain often performs poorly when it is applied or tested on opinionated documents from another domain, as we demonstrated through the example presented in Sect. 1. The reason is that words and even language constructs used in different domains for expressing opinions can be quite different. To make matters worse, the same word in one domain may have positive connotations, but in another domain may have negative ones; therefore, domain adaptation is needed. In the literature, different approaches related to the Multi-Domain sentiment analysis have been proposed. Briefly, two main categories may be identified: (i) the transfer of learned classifiers across different domains [3,4,58], and (ii) the use of propagation of labels through graph structures [21,59–61].

All approaches presented above are based on the use of statistical techniques for building sentiment models. The exploitation of semantic information is not taken into account. In this work, we proposed a first version of a semantic-based approach preserving the semantic relationships between the terms of each sentence in order to exploit them either for building the model and for estimating document polarity. The proposed approach, falling into the multi-domain sentiment analysis category, instead of using pre-determined polarity information associated with terms, it learns them directly from domain-specific documents. Such documents are used for training the models used by the system.

3   Network Description

Our approach can be described as follows: we process the text of a review word by word, using a Recurrent Neural Network, in what is called the *encoding* phase. The i-th state of the recurrent layer will summarize the text w.r.t. what has been seen up to the i-th word. So, the last state can be seen as a single distributed representation of the whole review. We exploit such representation to train the network to both estimate the property for the input review to belong to each of the domain in the training set *and* to predict the binary polarity.

In this section, after settling some basic notation and typographical conventions, the different components of the network will be described in detail.

3.1 Notation and Conventions

We used bold uppercase letters for matrices, say $\mathbf{W}$, and bold lowercase for vectors, say $\mathbf{b}$. When representing a vector explicitly in its components, we use square brackets and subscripts to indicate the position of each component within the vector, like in $\mathbf{x} = [x_1, \ldots, x_n]$. To represent a time series, we use the superscript between angle brackets for the components. So, if a $l \times h$ matrix is a time series of l vectors of size h, we write $\mathbf{H} = [\mathbf{h}^{\langle 1 \rangle}, \ldots, \mathbf{h}^{\langle l \rangle}]$. Similarly, if a vector represents a time series of scalar, we will write $\mathbf{x} = [x^{\langle 1 \rangle}, \ldots, x^{\langle n \rangle}]$. Apart from the set of real numbers $\mathbb{R}$, sets will be denoted with uppercase letters, like V. The number of elements of a finite set is written as $|V|$. Finally, we will indicate the set of parameters of our model with the capital Greek letter θ.

3.2 Input and Embedding Layer

A review is represented as a temporal sequence of words, as in $\mathbf{s} = [s^{\langle 1 \rangle}, \ldots, s^{\langle n \rangle} = $ <EOS>]. Conventionally, we append a special symbol <EOS> signaling the end of the sequence after the last word. Each word is mapped to a natural number called its *index*, representing the position of the word within the the the vocabulary V, i.e. the set of all the words in the input language—e.g. if the word chair is the 9th word in our vocabulary, we will have that $i_V(\text{chair}) = 9$. We can represent the review with a temporal sequence of natural numbers, the k-th of which represents the index of the k-th word in the review text: $\mathbf{i}_s = [i^{\langle 1 \rangle}, \ldots, i^{\langle n \rangle}] = [i_V(s^{\langle 1 \rangle}), \ldots, i_V(s^{\langle n \rangle}) = i_V(\text{<EOS>})]$. From now on, we will omit both the vocabulary subscript and the word argument and will write just $i^{\langle k \rangle}$ to indicate the k-th element of the indexes sequence.

We can pack the $|V|$ word vectors as rows of an *embedding matrix*, say $\mathbf{E}$, defined in $\mathbb{R}^{|V| \times d}$, where d is the dimension of the word vector, or *embedding size*, an hyper-parameter of the model. So, said $i^{\langle k \rangle}$ the index of the k-the word in the review, the corresponding word vector $\mathbf{x}^{\langle k \rangle}$ will be the $i^{\langle k \rangle}$-th row of the $\mathbf{E}$ matrix, writing:

$$\mathbf{x}^{\langle k \rangle} = \mathbf{E}[i^{\langle k \rangle}]. \tag{1}$$

The whole review can be represented as a sequence of word embedding vectors, $\mathbf{X} = [\mathbf{x}^{\langle 1 \rangle}, \ldots, \mathbf{x}^{\langle n \rangle}]$. The embedding matrix is previously learnt via the skip-gram model in a separate phase.

3.3 The Encoding Layer

With the term *encoding* we mean the act of capturing the whole review $\mathbf{s}$ into a single distributed vector $\mathbf{c} \in \mathbb{R}^h$, where h is an hyperparameter. Such operation is performed through a *recurrent* layer, where, at the k-th timestep, the output of the layer depends on the current input *and* on the activation at the previous step, as in:

$$\mathbf{h}^{\langle k \rangle} = g(\mathbf{x}^{\langle k \rangle}, \mathbf{h}^{\langle k-1 \rangle}; \theta_{enc}). \tag{2}$$

We use the term *activation* for the generic $\mathbf{h}^{\langle k \rangle}$ vector. The function g is the so called *cell function*, depending on a set of parameters θ_{enc} that are learnt during the training. The vector c is the activation of the last time step and is called the *encoder output*. We can write:

$$enc(\mathbf{s}) = \mathbf{c} = \mathbf{h}^{\langle n \rangle} = g(\mathbf{x}^{\langle n \rangle}, \mathbf{h}^{\langle n-1 \rangle}; \theta_{enc}). \tag{3}$$

Regarding the cell function of our encoder, we used the LSTM model. We won't give any detailed description of the mathematical model underlying such cell function, for which we refer to [62]. The main intuition about LSTM cells is that they are capable to maintain memory of the processed input across large time spans. This feature allows to partially overcome the problem of gradient vanishing (see [63]).

3.4   Output Layers and Loss Functions

The vector $\mathbf{c}$ resulting from the encoding phase, is then fed into an output layer, that allows to train the network jointly for the task of both domain and polarity identification. Such *dual* output layer is described below.

Domain identification. To identify the domain which the current input review belongs to, we feed the encoder output into a projection layer ending up into:

$$\mathbf{y} = softmax(\mathbf{W}_y \mathbf{c} + \mathbf{b}_y); \tag{4}$$

where $\mathbf{W}_y \in \mathbb{R}^{h \times |D|}$ and $\mathbf{b}_y \in \mathbb{R}^{|D|}$, said D the set of all the domains in the training set. The resulting vector $\mathbf{y}$ is a vector of dimension $|D|$, with one component per domain. The softmax operator compress the range of all the components of the vector into the $[0, 1]$ range and ensures that their value sums up to 1. In this way, the vector $\mathbf{y}$ can be seen as a probability distribution over the different domains: the value of the j-th component of the vector approximates the probability that the input review belongs to the j-th domain:

$$y_j = p(domain(\mathbf{s}) = j|\theta), \tag{5}$$

and the predicted domain is the one associated with the component with the maximum probability value, namely:

$$d_{pred} = \arg\max_j p(domain(\mathbf{s}) = j|\theta). \tag{6}$$

Said $\hat{\mathbf{y}}$ as the gold truth value for the domain identification of a given review, we use the categorical cross entropy between such gold truth and the predicted output as the loss function for the given example as:

$$\mathcal{L}_\mathbf{s} = -\sum_{j=1}^{|D|} \hat{y}_j log(y_j), \tag{7}$$

where $\hat{\mathbf{y}}$ is a one-hot vector, with all the values set to 0 but the one corresponding to the correct domain set to 1. The parameters of this layer are denoted with $\theta_y = [\mathbf{W}_y, \mathbf{b}_y]$ and will be learnt during the training process.

Polarity identification. For the decision about the polarity of the current input review, we use a slightly different output layer w.r.t. the one presented above. The output is still a vector of dimension $|D|$, one for each known domain, but each of its values range in $[-1, 1]$. Such vector is obtained via:

$$\mathbf{z} = tanh(\mathbf{W}_z\mathbf{c} + \mathbf{b}_z); \tag{8}$$

where $\mathbf{W}_z \in \mathbb{R}^{h \times |D|}$, $\mathbf{b}_z \in \mathbb{R}^{|D|}$, and $tanh(\cdot)$ is the hyperbolic tangent function. The j-th component of the vector represents the polarity score for the corresponding domain, namely:

$$z_j = p(polarity(\mathbf{s}) = j|\theta). \tag{9}$$

The value of the correct polarity is represented with another vector $\hat{\mathbf{z}}$ of $|D|$ dimensions, filled with 0 except for the position corresponding to the proper domain, where the value of the negative or positive label, namely $\{0; 1\}$ is set. As for the domain identification case, we use the categorical cross entropy as the loss function:

$$\mathcal{L}_\mathbf{s} = -\sum_{j=1}^{|D|} \hat{p}_j log(p_j). \tag{10}$$

The parameters of this layer that must be learned during the training phase are $\theta_z = [\mathbf{W}_z, \mathbf{b}_z]$.

The final polarity prediction is performed after a weighted sum with the different domain probabilities. The process is described in detail in Sect. 4.3.

4 Platform Architecture

Figure 1 shows the overall architecture of the proposed approach. This architecture has been entirely developed in Java with the support of the Deeplearning4j library [1] and it is composed by three main phases:

- Generation of Word vectors (Sect. 4.1): raw text, appropriately tokenized using the Stanford CoreNLP Toolkit, is provided as input to a 2-layers neural network implementing the skip-gram approach with the aim of generating word vectors.
- Learning of Sentiment Model (Sect. 4.2): word vectors are used for training a recurrent neural network implementing two output layers supporting the two classification tasks mentioned in the previous section: "polarity classification", where the word vector sequence is classified as positive or negative, and "domain classification" in which it is provided an overlap degree between the word vector sequence and each domain used for training the model.
- Computation of Document Polarity (Sect. 4.3): numeric data provided by the output layers are aggregated for inferring the overall polarity value of a text.

In the following subsections, we describe in more detail each phase by providing also the settings used for managing our data.

---

[1] https://deeplearning4j.org/.

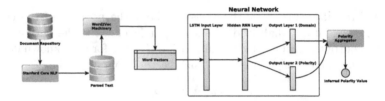

Fig. 1. Architecture of the proposed approach.

4.1   Generation of Word Vectors

The generation of the word vectors has been performed by applying the skip-gram algorithm on the raw natural language text extracted from the smaller version of the SNAP dataset [64]. The rationale behind the choice of this dataset focuses on three reasons:

- the dataset contains only opinion-based documents. This way, we are able to build word embeddings describing only opinion-based contexts.
- the dataset is multi-domain. Information contained into the generated word embeddings comes from specific domains, thus it is possible to evaluate how the proposed approach is general by testing the performance of the created model on test sets containing documents coming from the domains used for building the model or from other domains.
- the dataset is smaller with respect to other corpora used in the literature for building other word embeddings that are currently freely available, like the Google News ones. [2] Indeed, as introduced in Sect. 1, one of our goal is to demonstrate how we can leverage the use of dedicated resources for generating word embeddings, instead of corpora's size, for improving the effectiveness of classification systems.

These three points represent the main original contributions of this work, in particular the aspect of considering only opinion-based information for generating word embeddings. While embeddings currently available are created from big corpora of general purpose texts (like news archives or Wikipedia pages), ours are generated by using a smaller corpus containing documents strongly related to the problem that the model will be thought for. On the one hand, this aspect may be considered a limitation of the proposed solution due to the requirement of training a new model in case of problem change. However, on the other hand, the usage of dedicated resources would lead to the construction of more effective models.

Word embeddings have been generated by the Word2Vec implementation integrated into the Deeplearning4j library. The algorithm has been set up with the following parameters: the size of the vector to 64, the size of the window used as input of the skip-gram algorithm to 5, and the minimum word frequency was set to 1. The reason for which we kept the minimum word frequency set to 1 is

---

[2] https://github.com/mmihaltz/word2vec-GoogleNews-vectors.

to avoid the loss of rare but important words that can occur in domain specific documents like the ones contained in the Blitzer dataset.

4.2   Learning of The Sentiment Model

The sentiment model is built by starting from the word embeddings generated during the previous phase. In Sect. 3.4, we already introduced both the tasks performed by our model and we generally refer to Sect. 3 for the mathematical details of our approach. Here, we summarize the main steps performed for building our model, from the conversion of the input sentences to the strategy used for inferring the overall polarity of a document.

The first step consists in converting each textual sentence contained within the dataset into the corresponding numerical matrix $\mathbf{S}$ where we have in each row the word vector representing a single word of the sentence, and in each column an embedding feature. Given a sentence s, we extract all tokens t_i, with $i \in [0, n]$, and we replace each t_i with the corresponding embedding $\mathbf{w}$. During the conversion of each word in its corresponding embedding, if such embedding is not found, the word is discarded. At the end of this step, each sentence contained in the training set is converted in a matrix $\mathbf{S} = [\mathbf{w}^{\langle 1 \rangle}, \ldots, \mathbf{w}^{\langle n \rangle}]$.

Before giving all matrices as input to the neural network, we need to include both padding and masking vectors in order to train our model correctly. Padding and masking allows us to support different training situations depending on the number of the input vectors and on the number of predictions that the network has to provide at each time step. In our scenario, we work in a many-to-one situation where our neural network has to provide one prediction (sentence polarity and domain overlap) as result of the analysis of many input vectors (word embeddings).

Padding vectors are required because we have to deal with the different length of sentences. Indeed, the neural network needs to know the number of time steps that the input layer has to import. This problem is solved by including, if necessary, into each matrix $\mathbf{S}_k$, with $k \in [0, z]$ and z the number of sentences contained in the training set, null word vectors that are used for filling empty word's slots. These null vectors are accompanied by a further vector telling to the neural network if data contained in a specific positions has to be considered as an informative embedding or not.

A final note concerns the back propagation of the error. Training recurrent neural networks can be quite computationally demanding in cases when each training instance is composed by many time steps. A possible optimization is the use of truncated back propagation through time (BPTT) that was developed for reducing the computational complexity of each parameter update in a recurrent neural network. On the one hand, this strategy allows to reduce the time needed for training our model. However, on the other hand, there is the risk of not flowing backward the gradients for the full unrolled network. This prevents the full update of all network parameters. For this reason, even if we work with recurrent neural networks, we decided to do not implement a BPTT approach but to use the default backpropagation implemented into the DL4J library.

Concerning information about network structure, the input layer was composed by 64 neurons (i.e. embedding vector size), the hidden RNN layer was composed by 128 nodes, and the output layers contained 20 nodes each. The network has been trained by using the Stochastic Gradient Descent with 1000 epochs and a learning rate of 0.002.

4.3   Computation of Document Polarity

The operation of computing the entire polarity of a document is performed by combining the numeric data provided by the two output layers of the network. As explained earlier, one of the assumptions of the proposed approach is to exploit possible linguistic overlaps between different domains for compensating missing knowledge from the training set. From here, the intuition of using two output layers: the first one for computing the belonging degree between a sentence and each domain, and the second one for computing the polarity of the document with respect to each domain. Given $\mathbf{y}$ the vector of dimension $|D|$ containing the belonging degree of a sentence to each domain, and $\mathbf{z}$, the vector of dimension $|D|$ containing the polarity score for each domain, the overall predicted polarity of a document T is computed by scaling the dot product between the two vectors by their dimension, as follow:

$$p_T = \frac{\mathbf{y} \cdot \mathbf{z}}{|D|} \tag{11}$$

5   Evaluation

The NeuroSent approach discussed in Sect. 3 and the platform presented in Sect. 4 have been evaluated by adopting the Dranziera protocol [65]. Here, we describe the implemented validation procedure and we discuss the obtained results.

5.1   Evaluation Procedure

The validation procedure leverage on a five-fold cross evaluation setting in order to validate the robustness of the proposed solution. The approach has been compared with seven baselines:

- Support Vector Machine (SVM): classification was run with a linear kernel type by using the Libsvm [66]. Libsvm uses a sparse format so that zero values do not need to be captured for training files. This can cause training time to be longer, but keeps Libsvm flexible for sparse cases.
- Naive Bayes (NB) and Maximum Entropy (ME): the MALLET: MAchine Learning for LanguagE Toolkit [67] was used for classification by using both Naive Bayes and Maximum Entropy algorithms. For the experiments conducted in our evaluation, the Maximum Entropy classification has been performed by using a Gaussian prior variance of 1.0.

- Domain Belonging Polarity (DBP): we computed the text polarity by using only information of the domain that a document belongs to. This means that the linguistic overlap between domains has not been considered.
- Domain Detection Polarity (DDP): we computed the text polarity by using only information of the domain guessed as the most appropriate for the document that has to be evaluated. This means that the similarity between text content and domain is preferred with respect to the domain used for tagging the text.
- Convolutional Neural Network [68] (CNN): we compared our architecture with a classic CNN. Models have been trained with the embeddings created from the Blitzer dataset.
- Google Word Embeddings (GWE): we trained our models by using the pre-trained Google-News word vectors [3] instead of the ones created from the Blitzer dataset. The goal of this comparison is to show how the embeddings created by using a smaller, but more opinion-oriented, dataset may lead to better results with respect to the use of bigger, but general, embeddings.

For each baseline, we measured the overall accuracy, the precision and recall averaged over the two classes (positive and negative) and the F1-score. In the end, we reported their averages together with the standard deviation measured over the five folds.

The same baselines have been used to evaluate the model against the OMD test sets. In this case, the DBP baseline has not been applied due to the mismatch between the domains used for building the model and the ones contained in the test sets. Each OMD test set has been applied to all five models built, and the scores averaged.

5.2   Results on In-Vitro Dataset

Here, we show the results of the evaluation campaign conducted to validate the presented approach. Tables 1 and 2 present a summary of the performance obtained by NeuroSent and by the seven baselines on IMD and OMD, respectively. The first column contains the name of the approach, the second, third, and fourth contain the average precision, recall and F1 score computed over all domains, while the fifth contains the average standard deviation computed on the F1 score during the cross-fold validation. Finally, the sixth and seventh columns contain the minimum and the maximum F1 score measured during the evaluation.

Table 1 shows the results obtained on the domains contained in the Dranziera dataset; while in Table 2, we presented the results obtained by testing our approach on documents coming from domains that are not contained in the Dranziera dataset.

By considering the overall results obtained on the IMD and reported in Table 1, we may observe how NeuroSent outperforms the considered baselines.

---

[3] https://github.com/mmihaltz/word2vec-GoogleNews-vectors.

Table 1. Comparison between the results obtained by the baselines and the ones obtained by NeuroSent on the Dranziera dataset.

Approach	Avg. Precision	Avg. Recall	Avg. F1	Avg. Deviation
Support Vector Machine	0.6890	0.7097	0.6987	0.0119
Naive-Bayes	0.6956	0.6915	0.6929	0.0062
Maximum Entropy	0.7073	0.7085	0.7074	0.0098
Domain Belonging Polarity	0.7108	0.7331	0.7218	0.0228
Domain Detection Polarity	0.6731	0.7546	0.7115	0.0384
CNN Architecture	0.8037	0.7727	0.7879	0.0516
Google Word Embeddings	0.8008	0.7921	0.7964	**0.0042**
NeuroSent	**0.8687**	**0.8563**	**0.8625**	0.0098

The measured F1 scores (average, minimum, and maximum) are higher of around 6% with respect to the GWE and CNN baselines and of more than 8–10% with respect to the others. This first consideration demonstrated the superiority of the use of word embeddings with respect to strategies that do not implement this technique. The poor performance of the classic machine learning approaches: SVM, NB, and ME, are surprising. A more in depth analysis of the results obtained by these three baselines shown how these approaches failed in classifying long reviews containing many opinion-based sentences.

A similar rank can be observed by analyzing the precision and recall values, where the differences between NeuroSent and the baselines remains more or less the same. Concerning the stability of the algorithm, we can notice that the exploitation of the domain information leads to a higher standard deviation. In both IMD and OMD test sets, the domain-based approaches registered a higher standard deviation with respect to the SVM, NB, and ME baselines.

Table 2. Comparison between the results obtained by the baselines and the ones obtained by NeuroSent on the domains not contained in the Dranziera dataset.

Approach	Avg. Precision	Avg. Recall	Avg. F1	Avg. Deviation
Support Vector Machine	0.6507	0.6437	0.6571	0.0090
Naive-Bayes	0.6435	0.6459	0.6260	**0.0023**
Maximum Entropy	0.6524	0.6475	0.6501	0.0038
Domain Belonging Polarity	-	-	-	-
Domain Detection Polarity	0.6609	0.6931	0.6766	0.0096
CNN Architecture	0.7848	0.8120	0.7982	0.0457
Google Word Embeddings	0.7930	0.7864	0.7896	0.0078
NeuroSent	**0.8509**	**0.8478**	**0.8493**	0.0109

The second overall evaluation concerns the analysis of the results obtained on the set of OMD. Results are shown in Table 2. The first thing that we may observe is how the effectiveness obtained by the proposed system is very close to the one obtained in the IMD evaluation. Indeed, the difference between the two F1 averages is less than 1%. This aspect remarked the capability of the proposed approach of working in a cross-domain environment and of exploiting the domain linguistic overlaps for estimating the overall document polarity.

Thus, we may state that the exploitation of domain linguistic overlaps is a suitable solution for compensating the possible lack of knowledge when limited training sets are used for building opinion models.

Finally, we want to report some considerations about the efficiency of the platform. The time required for building the entire sentiment model was around 8.5 hours (single core equivalent) on a server equipped with a double Xeon X5650 and 32 Gb of RAM. While, during the testing session, on the same machine, the computation of a single document polarity required an average of 658 ms. This result supports the possible implementation of the polarity computation component as a real-time service due to the low time required for computing document polarity.

5.3   The NeuroSent System at ESWC-2018 SSA Challenge Task #1

The system participated to the Task #1 of the Semantic Sentiment Analysis Challenge co-located with ESWC 2016. Table 3 shows the results of Task #1. We may observe that the system ranked third, not far from the best performer. This result confirm the viability of the implemented approach for future implementation after a study of the main error scenarios in which the fuzzy-based algorithm performed poorly.

Table 3. Results obtained by the participant to Task #1.

System	F-Measure
Guangyuan et al.	0.9643
Atzori et al.	0.9561
Federici et al.	0.9356
Dragoni et al.	**0.9228**
Dragoni et al.	0.8823
Rexha et al.	0.8743
Indurthi et al.	0.8203
Petrucci et al.	0.7153
Chaudhuri et al.	0.5243

6 Conclusion and Future Work

In this paper, we presented NeuroSent: a tool for multi-domain sentiment analysis exploiting linguistic overlaps between domains for inferring document polarity. The tool implements a deep learning architecture using distributed vectors to represent words.

Models are built by using a recurrent neural network trained with information extracted from the Blitzer dataset, a small but opinion-oriented collection of user-generated reviews. The inference of the overall polarity of a document is performed by combining belonging degrees of a document to each domain and domain-specific polarities scores computed by the model.

The performance of the system has been evaluated by applying the Dranziera protocol. Results shown the effectiveness of the proposed approach with respect to the baselines demonstrating its viability. Moreover, the protocol used for the evaluation enables an easy reproducibility of the experiments and the comparison with other systems.

Future work will focus on two main directions: (i) the integration of knowledge embeddings and (ii) the injection of fuzzy logic for representing uncertainty associated with word embeddings. Concerning the integration of knowledge embeddings, we want to build embeddings describing complex sentiment patterns instead of single terms (or concepts). This way, the system will be able to learn more complex linguistic constructs with the aim of improving the overall classification effectiveness. While, concerning the injection of fuzzy logic, our intent is to represent each embedding's feature by using fuzzy sets. On the one hand, this solution allows to manage uncertainty associated with each feature; however, on the other hand, the complexity of the overall architecture will increase. The challenge will focus on finding an efficient way to manage all such information. Some effort will be also focused on the creation of domain-dependent sentiment lexicons and on validating their effectiveness with respect to general purpose ones.

Finally, we foresee the integration of a concept extraction approach in order to provide the system with further semantic capabilities, for example the extraction of finer-grained information, that can be used during the model construction.

References

1. Cambria, E., White, B.: Jumping NLP curves: a review of natural language processing research [review article]. IEEE Comp. Int. Mag. **9**(2), 48–57 (2014)
2. Pang, B., Lee, L., Vaithyanathan, S.: Thumbs up? sentiment classification using machine learning techniques. In: Proceedings of EMNLP, Philadelphia, Association for Computational Linguistics, pp. 79–86 (July 2002)
3. Blitzer, J., Dredze, M., Pereira, F.: Biographies, bollywood, boom-boxes and blenders: domain adaptation for sentiment classification. In: Carroll, J.A., van den Bosch, A., Zaenen, A. (eds.) ACL 2007, Proceedings of the 45th Annual Meeting of the Association for Computational Linguistics, June 23–30, 2007, Prague, Czech Republic. The Association for Computational Linguistics (2007)

4. Pan, S.J., Ni, X., Sun, J.T., Yang, Q., Chen, Z.: Cross-domain sentiment classification via spectral feature alignment. In: Rappa, M., Jones, P., Freire, J., Chakrabarti, S. (eds.) Proceedings of the 19th International Conference on World Wide Web, WWW 2010, Raleigh, North Carolina, USA, April 26–30, 2010, pp. 751–760. ACM (2010)
5. Liu, B., Zhang, L.: A survey of opinion mining and sentiment analysis. In: Aggarwal, C.C., Zhai, C.X. (eds.) Mining Text Data, pp. 415–463. Springer, Berlin (2012)
6. Pang, B., Lee, L.: A sentimental education: sentiment analysis using subjectivity summarization based on minimum cuts. In: Scott, D., Daelemans, W., Walker, M.A. (eds.) Proceedings of the 42nd Annual Meeting of the Association for Computational Linguistics, 21–26 July, 2004, Barcelona, Spain., pp. 271–278. ACL (2004)
7. Dave, K., Lawrence, S., Pennock, D.M.: Mining the peanut gallery: opinion extraction and semantic classification of product reviews. In: WWW, pp. 519–528 (2003)
8. Paltoglou, G., Thelwall, M.: A study of information retrieval weighting schemes for sentiment analysis. In: ACL, pp. 1386–1395 (2010)
9. Tan, S., Wang, Y., Cheng, X.: Combining learn-based and lexicon-based techniques for sentiment detection without using labeled examples. In: Myaeng, S., Oard, D.W., Sebastiani, F., Chua, T., Leong, M. (eds.) Proceedings of the 31st Annual International ACM SIGIR Conference on Research and Development in Information Retrieval, SIGIR 2008, Singapore, July 20–24, 2008, pp. 743–744. ACM (2008)
10. Qiu, L., Zhang, W., Hu, C., Zhao, K.: SELC: a self-supervised model for sentiment classification. In: Cheung, D.W., Song, I., Chu, W.W., Hu, X., Lin, J.J. (eds.) Proceedings of the 18th ACM Conference on Information and Knowledge Management, CIKM 2009, Hong Kong, China, November 2–6, 2009, pp. 929–936. ACM (2009)
11. Melville, P., Gryc, W., Lawrence, R.D.: Sentiment analysis of blogs by combining lexical knowledge with text classification. In: KDD, pp. 1275–1284 (2009)
12. Taboada, M., Brooke, J., Tofiloski, M., Voll, K.D., Stede, M.: Lexicon-based methods for sentiment analysis. Comput. Linguist. **37**(2), 267–307 (2011)
13. Somasundaran, S.: Discourse-level relations for Opinion Analysis. Ph.D. thesis, University of Pittsburgh (2010)
14. Wang, H., Zhou, G.: Topic-driven multi-document summarization. In: IALP, pp. 195–198 (2010)
15. Dragoni, M.: Shellfbk: An information retrieval-based system for multi-domain sentiment analysis. In: Proceedings of the 9th International Workshop on Semantic Evaluation. SemEval '2015, Denver, Colorado, pp. 502–509. Association for Computational Linguistics (June 2015)
16. Petrucci, G., Dragoni, M.: An information retrieval-based system for multi-domain sentiment analysis. In: Gandon, F., Cabrio, E., Stankovic, M., Zimmermann, A. (eds.) SemWebEval 2015. CCIS, vol. 548, pp. 234–243. Springer, Cham (2015). https://doi.org/10.1007/978-3-319-25518-7_20
17. Rexha, A., Kröll, M., Dragoni, M., Kern, R.: Exploiting propositions for opinion mining. In: Sack, H., Dietze, S., Tordai, A., Lange, C. (eds.) SemWebEval 2016. CCIS, vol. 641, pp. 121–125. Springer, Cham (2016). https://doi.org/10.1007/978-3-319-46565-4_9
18. Federici, M., Dragoni, M.: A knowledge-based approach for aspect-based opinion mining. In: Sack, H., Dietze, S., Tordai, A., Lange, C. (eds.) SemWebEval 2016. CCIS, vol. 641, pp. 141–152. Springer, Cham (2016). https://doi.org/10.1007/978-3-319-46565-4_11

19. Rexha, A., Kröll, M., Dragoni, M., Kern, R.: Opinion mining with a clause-based approach. In: Dragoni, M., Solanki, M., Blomqvist, E. (eds.) SemWebEval 2017. CCIS, vol. 769, pp. 166–175. Springer, Cham (2017). https://doi.org/10.1007/978-3-319-69146-6_15

20. Federici, M., Dragoni, M.: Aspect-based opinion mining using knowledge bases. In: Dragoni, M., Solanki, M., Blomqvist, E. (eds.) SemWebEval 2017. CCIS, vol. 769, pp. 133–147. Springer, Cham (2017). https://doi.org/10.1007/978-3-319-69146-6_13

21. Dragoni, M., Tettamanzi, A.G.B., da Costa Pereira, C.: Propagating and aggregating fuzzy polarities for concept-level sentiment analysis. Cogn. Comput. **7**(2), 186–197 (2015)

22. Dragoni, M., Tettamanzi, A.G.B., da Costa Pereira, C.: A fuzzy system for concept-level sentiment analysis. In: Presutti, V., Stankovic, M., Cambria, E., Cantador, I., Di Iorio, A., Di Noia, T., Lange, C., Reforgiato Recupero, D., Tordai, A. (eds.) SemWebEval 2014. CCIS, vol. 475, pp. 21–27. Springer, Cham (2014). https://doi.org/10.1007/978-3-319-12024-9_2

23. Petrucci, G., Dragoni, M.: The IRMUDOSA system at ESWC-2016 challenge on semantic sentiment analysis. In: Sack, H., Dietze, S., Tordai, A., Lange, C. (eds.) SemWebEval 2016. CCIS, vol. 641, pp. 126–140. Springer, Cham (2016). https://doi.org/10.1007/978-3-319-46565-4_10

24. Dragoni, M., Petrucci, G.: A fuzzy-based strategy for multi-domain sentiment analysis. Int. J. Approx. Reason. **93**, 59–73 (2018)

25. Petrucci, G., Dragoni, M.: The IRMUDOSA system at ESWC-2017 challenge on semantic sentiment analysis. In: Dragoni, M., Solanki, M., Blomqvist, E. (eds.) SemWebEval 2017. CCIS, vol. 769, pp. 148–165. Springer, Cham (2017). https://doi.org/10.1007/978-3-319-69146-6_14

26. da Costa Pereira, C., Dragoni, M., Pasi, G.: A prioritized "and" aggregation operator for multidimensional relevance assessment. In: Serra, R., Cucchiara, R. (eds.) AI*IA 2009. LNCS (LNAI), vol. 5883, pp. 72–81. Springer, Heidelberg (2009). https://doi.org/10.1007/978-3-642-10291-2_8

27. Federici, M., Dragoni, M.: Towards unsupervised approaches for aspects extraction. In: Dragoni, M., Recupero, D.R., Denecke, K., Deng, Y., Declerck, T. (eds.) Joint Proceedings of the 2th Workshop on Emotions, Modality, Sentiment Analysis and the Semantic Web and the 1st International Workshop on Extraction and Processing of Rich Semantics from Medical Texts co-located with ESWC 2016, Heraklion, Greece, May 29, 2016. Volume 1613 of CEUR Workshop Proceedings (2016). www.ceur-ws.org

28. Federici, M., Dragoni, M.: A branching strategy for unsupervised aspect-based sentiment analysis. In: Dragoni, M., Recupero, D.R., (eds.) Proceedings of the 3rd International Workshop at ESWC on Emotions, Modality, Sentiment Analysis and the Semantic Web co-located with 14th ESWC 2017, Portroz, Slovenia, May 28, 2017. Volume 1874 of CEUR Workshop Proceedings (2017). www.ceur-ws.org

29. Riloff, E., Patwardhan, S., Wiebe, J.: Feature subsumption for opinion analysis. In: Jurafsky, D., Gaussier, É. (eds.) EMNLP 2007, Proceedings of the 2006 Conference on Empirical Methods in Natural Language Processing, 22–23 July 2006, Sydney, Australia, pp. 440–448. ACL (2006)

30. Wilson, T., Wiebe, J., Hwa, R.: Recognizing strong and weak opinion clauses. Comput. Intell. **22**(2), 73–99 (2006)

31. Palmero Aprosio, A., Corcoglioniti, F., Dragoni, M., Rospocher, M.: Supervised opinion frames detection with RAID. In: Gandon, F., Cabrio, E., Stankovic, M.,

Zimmermann, A. (eds.) SemWebEval 2015. CCIS, vol. 548, pp. 251–263. Springer, Cham (2015). https://doi.org/10.1007/978-3-319-25518-7_22

32. Hatzivassiloglou, V., Wiebe, J.: Effects of adjective orientation and gradability on sentence subjectivity. In: COLING 2000, 18th International Conference on Computational Linguistics, Proceedings of the Conference, 2 Volumes, July 31 - August 4, 2000, Universität des Saarlandes, Saarbrücken, Germany, pp. 299–305. Morgan Kaufmann (2000)

33. Kim, S., Hovy, E.H.: Crystal: Analyzing predictive opinions on the web. In: Eisner, J. (ed.) EMNLP-CoNLL 2007, Proceedings of the 2007 Joint Conference on Empirical Methods in Natural Language Processing and Computational Natural Language Learning, June 28–30, 2007, Prague, Czech Republic, pp. 1056–1064. ACL (2007)

34. Rexha, A., Kröll, M., Dragoni, M., Kern, R.: Polarity classification for target phrases in tweets: a Word2Vec approach. In: Sack, H., Rizzo, G., Steinmetz, N., Mladenić, D., Auer, S., Lange, C. (eds.) ESWC 2016. LNCS, vol. 9989, pp. 217–223. Springer, Cham (2016). https://doi.org/10.1007/978-3-319-47602-5_40

35. Rexha, A., Kröll, M., Kern, R., Dragoni, M.: An embedding approach for microblog polarity classification. In: Dragoni, M., Recupero, D.R. (eds.) Proceedings of the 3rd International Workshop on Emotions, Modality, Sentiment Analysis and the Semantic Web co-located with 14th ESWC 2017, Portroz, Slovenia, May 28, 2017. Volume 1874 of CEUR Workshop Proceedings (2017). www.ceur-ws.org

36. Recupero, D.R., Dragoni, M., Presutti, V.: ESWC 15 challenge on concept-level sentiment analysis. In: Gandon, F., Cabrio, E., Stankovic, M., Zimmermann, A. (eds.) SemWebEval 2015. CCIS, vol. 548, pp. 211–222. Springer, Cham (2015). https://doi.org/10.1007/978-3-319-25518-7_18

37. Dragoni, M., Reforgiato Recupero, D.: Challenge on fine-grained sentiment analysis within ESWC2016. In: Sack, H., Dietze, S., Tordai, A., Lange, C. (eds.) SemWebEval 2016. CCIS, vol. 641, pp. 79–94. Springer, Cham (2016). https://doi.org/10.1007/978-3-319-46565-4_6

38. Dragoni, M., Solanki, M., Blomqvist, E. (eds.): SemWebEval 2017. CCIS, vol. 769. Springer, Cham (2017). https://doi.org/10.1007/978-3-319-69146-6

39. Jakob, N., Gurevych, I.: Extracting opinion targets in a single and cross-domain setting with conditional random fields. In: Proceedings of the 2010 Conference on Empirical Methods in Natural Language Processing, EMNLP 2010, 9–11 October 2010, MIT Stata Center, Massachusetts, USA, A meeting of SIGDAT, a Special Interest Group of the ACL, pp. 1035–1045. ACL (2010)

40. Jin, W., Ho, H.H., Srihari, R.K.: OpinionMiner: a novel machine learning system for web opinion mining and extraction. In: IV, J.F.E., Fogelman-Soulié, F., Flach, P.A., Zaki, M.J. (eds.) Proceedings of the 15th ACM SIGKDD International Conference on Knowledge Discovery and Data Mining, Paris, France, June 28 - July 1, 2009, pp. 1195–1204. ACM (2009)

41. Liu, B., Hu, M., Cheng, J.: Opinion observer: analyzing and comparing opinions on the web. In: Ellis, A., Hagino, T. (eds.) Proceedings of the 14th International Conference on World Wide Web, WWW 2005, Chiba, Japan, May 10–14, 2005, pp. 342–351. ACM (2005)

42. Wu, Y., Zhang, Q., Huang, X., Wu, L.: Phrase dependency parsing for opinion mining. In: Proceedings of the 2009 Conference on Empirical Methods in Natural Language Processing, EMNLP 2009, 6–7 August 2009, Singapore, A meeting of SIGDAT, a Special Interest Group of the ACL, pp. 1533–1541. ACL (2009)

43. Su, Q., Xu, X., Guo, H., Guo, Z., Wu, X., Zhang, X., Swen, B., Su, Z.: Hidden sentiment association in Chinese web opinion mining. In: Huai, J., Chen, R., Hon, H., Liu, Y., Ma, W., Tomkins, A., Zhang, X. (eds.) Proceedings of the 17th International Conference on World Wide Web, WWW 2008, Beijing, China, April 21–25, 2008, pp. 959–968. ACM (2008)

44. Dragoni, M.: NEUROSENT-PDI at semeval-2018 task 1: Leveraging a multi-domain sentiment model for inferring polarity in micro-blog text. In: Apidianaki, M., Mohammad, S.M., May, J., Shutova, E., Bethard, S., Carpuat, M. (eds.) Proceedings of The 12th International Workshop on Semantic Evaluation, SemEval@NAACL-HLT, New Orleans, Louisiana, June 5–6, 2018, pp. 102–108. Association for Computational Linguistics (2018)

45. Dragoni, M.: NEUROSENT-PDI at semeval-2018 task 3: understanding irony in social networks through a multi-domain sentiment model. In: Apidianaki, M., Mohammad, S.M., May, J., Shutova, E., Bethard, S., Carpuat, M. (eds.) Proceedings of The 12th International Workshop on Semantic Evaluation, SemEval@NAACL-HLT, New Orleans, Louisiana, June 5–6, 2018, pp. 512–519. Association for Computational Linguistics (2018)

46. Dragoni, M., Azzini, A., Tettamanzi, A.G.B.: A novel similarity-based crossover for artificial neural network evolution. In: Schaefer, R., Cotta, C., Kołodziej, J., Rudolph, G. (eds.) PPSN 2010. LNCS, vol. 6238, pp. 344–353. Springer, Heidelberg (2010). https://doi.org/10.1007/978-3-642-15844-5_35

47. Qiu, G., Liu, B., Bu, J., Chen, C.: Opinion word expansion and target extraction through double propagation. Comput. Linguist. 37(1), 9–27 (2011)

48. Dragoni, M., da Costa Pereira, C., Tettamanzi, A.G.B., Villata, S.: Combining argumentation and aspect-based opinion mining: the smack system. AI Commun. 31(1), 75–95 (2018)

49. Dragoni, M.: A three-phase approach for exploiting opinion mining in computational advertising. IEEE Intell. Syst. 32(3), 21–27 (2017)

50. Dragoni, M., Petrucci, G.: A neural word embeddings approach for multi-domain sentiment analysis. IEEE Trans. Affect. Comput. 8(4), 457–470 (2017)

51. Dragoni, M.: Computational advertising in social networks: an opinion mining-based approach. In: Haddad, H.M., Wainwright, R.L., Chbeir, R. (eds.) Proceedings of the 33rd Annual ACM Symposium on Applied Computing, SAC 2018, Pau, France, April 09–13, 2018, pp. 1798–1804. ACM (2018)

52. Barbosa, L., Feng, J.: Robust sentiment detection on twitter from biased and noisy data. In: Huang, C., Jurafsky, D. (eds.) COLING 2010, 23rd International Conference on Computational Linguistics, Posters Volume, 23–27 August 2010, Beijing, China, pp. 36–44. Chinese Information Processing Society of China (August 2010)

53. Bermingham, A., Smeaton, A.F.: Classifying sentiment in microblogs: is brevity an advantage? In: CIKM, pp. 1833–1836. (2010)

54. Go, A., Bhayani, R., Huang, L.: Twitter sentiment classification using distant supervision. CS224N Project Report, Standford University (2009)

55. Cambria, E., Hussain, A.: Sentic computing: a common-sense-based framework for concept-level sentiment analysis (2015)

56. Cambria, E., Hussain, A.: Sentic album: content-, concept-, and context-based online personal photo management system. Cogn. Comput. 4(4), 477–496 (2012)

57. Wang, Q.F., Cambria, E., Liu, C.L., Hussain, A.: Common sense knowledge for handwritten Chinese recognition. Cogn. Comput. 5(2), 234–242 (2013)

58. Yoshida, Y., Hirao, T., Iwata, T., Nagata, M., Matsumoto, Y.: Transfer learning for multiple-domain sentiment analysis–identifying domain dependent/independent word polarity. In: AAAI, pp. 1286–1291 (2011)

59. Ponomareva, N., Thelwall, M.: Semi-supervised vs. cross-domain graphs for sentiment analysis. In: Angelova, G., Bontcheva, K., Mitkov, R. (eds.) Recent Advances in Natural Language Processing, RANLP 2013, 9–11 September, 2013, Hissar, Bulgaria, pp. 571–578. RANLP 2013 Organising Committee/ACL (2013)
60. Huang, S., Niu, Z., Shi, C.: Automatic construction of domain-specific sentiment lexicon based on constrained label propagation. Knowl.-Based Syst. **56**, 191–200 (2014)
61. Dragoni, M., da Costa Pereira, C., Tettamanzi, A.G.B., Villata, S.: Smack: an argumentation framework for opinion mining. In: Kambhampati, S. (ed.) Proceedings of the Twenty-Fifth International Joint Conference on Artificial Intelligence, IJCAI 2016, New York, NY, USA, 9–15 July 2016, pp. 4242–4243. IJCAI/AAAI Press (2016)
62. Hochreiter, S., Schmidhuber, J.: Long short-term memory. Neural Comput. **9**(8), 1735–1780 (1997)
63. Hochreiter, S.: Untersuchungen zu dynamischen neuronalen netzen. Diploma , Technische Universität München(1991)
64. McAuley, J.J., Leskovec, J.: Hidden factors and hidden topics: understanding rating dimensions with review text. In: Yang, Q., King, I., Li, Q., Pu, P., Karypis, G. (eds.) Seventh ACM Conference on Recommender Systems, RecSys '13, Hong Kong, China, October 12–16, 2013, pp. 165–172. ACM (2013)
65. Dragoni, M., Tettamanzi, A.G.B., da Costa Pereira, C.: DRANZIERA: an evaluation protocol for multi-domain opinion mining. In: Calzolari, N., et al. (eds.) Proceedings of the Tenth International Conference on Language Resources and Evaluation (LREC 2016), Paris, France, European Language Resources Association (ELRA) (May 2016)
66. Chang, C.C., Lin, C.J.: Libsvm: a library for support vector machines. ACM TIST **2**(3), 27:1–27:27 (2011)
67. McCallum, A.K.: Mallet: a machine learning for language toolkit (2002). http://mallet.cs.umass.edu
68. Chaturvedi, I., Cambria, E., Vilares, D.: Lyapunov filtering of objectivity for spanish sentiment model. In: 2016 International Joint Conference on Neural Networks, IJCNN 2016, Vancouver, BC, Canada, July 24–29, 2016, pp. 4474–4481. IEEE (2016)

The FeatureSent System at ESWC-2018 Challenge on Semantic Sentiment Analysis

Mauro Dragoni[(✉)]

Fondazione Bruno Kessler, Trento, Italy
dragoni@fbk.eu

Abstract. The approach described in this paper explores the use of semantic structured representation of sentences extracted from texts for multi-domain sentiment analysis purposes. The presented algorithm is built upon a domain-based supervised approach using index-like structured for representing information extracted from text. The algorithm extracts dependency parse relationships from the sentences containing in a training set. Then, such relationships are aggregated in a semantic structured together with either polarity and domain information. Such information is exploited in order to have a more fine-grained representation of the learned sentiment information. When the polarity of a new text has to be computed, such a text is converted in the same semantic representation that is used (i) for detecting the domain to which the text belongs to, and then (ii), once the domain is assigned to the text, the polarity is extracted from the index-like structure. First experiments performed by using the Blitzer dataset for training the system demonstrated the feasibility of the proposed approach.

1 Introduction

Sentiment analysis is a natural language processing task whose aim is to classify documents according to the opinion (polarity) they express on a given subject [1]. Generally speaking, sentiment analysis aims at determining the attitude of a speaker or a writer with respect to a topic or the overall tonality of a document. This task has created a considerable interest due to its wide applications. In recent years, the exponential increase of the Web for exchanging public opinions about events, facts, products, etc., has led to an extensive usage of sentiment analysis approaches, especially for marketing purposes.

By formalizing the sentiment analysis problem, a "sentiment" or "opinion" has been defined by [2] as a quintuple:

$$\langle o_j, f_{jk}, so_{ijkl}, h_i, t_l \rangle, \tag{1}$$

where o_j is a target object, f_{jk} is a feature of the object o_j, so_{ijkl} is the sentiment value of the opinion of the opinion holder h_i on feature f_{jk} of object o_j

© Springer Nature Switzerland AG 2018
D. Buscaldi et al. (Eds.): SemWebEval 2018, CCIS 927, pp. 216–231, 2018.
https://doi.org/10.1007/978-3-030-00072-1_17

at time t_l. The value of so_{ijkl} can be positive (by denoting a state of happiness, bliss, or satisfaction), negative (by denoting a state of sorrow, dejection, or disappointment), or neutral (it is not possible to denote any particular sentiment), or a more granular rating. The term h_i encodes the opinion holder, and t_l is the time when the opinion is expressed.

Such an analysis, may be *document-based*, where the positive, negative, or neutral sentiment is assigned to the entire document content; or *sentence-based* where individual sentences are analyzed separately and classified according to the different polarity values. In the latter case, it is often desirable to find with a high precision the entity attributes towards which the detected sentiment is directed. Based on the scenario in which the opinion is needed, the use of a document-based analysis is preferred with respect to a sentence-based one, and vice versa. In this work, we want to extract the general opinion of an entire document; therefore, our approach relies on a document-based analysis.

A further aspect that it is important to take into account is that, in the classic sentiment analysis problem, the polarity of each document term is considered independently by the domain which the document belongs to. We illustrate the intuition behind domain specific term polarity by considering the following example:

1. The sideboard is **small** and it is not able to contain a lot of stuff.
2. The **small** dimensions of this decoder allow to move it easily.

In these two sentences the adjective "small" is used in two different domains. In the first sentence, we considered the Furnishings domain and, within it, the polarity of the adjective "small" is, for sure, "negative" because it highlights an issue of the described item. On the other hand, in the second sentence, where we considered the Electronics domain, the polarity of such an adjective may be considered "positive". First attempts exploring how term polarity is conditioned by domain is presented in [3].

Unlike the approaches already discussed in the literature (presented in Sect. 2), we address the multi-domain sentiment analysis problem from a different perspective. Firstly, we extract semantic and linguistic relationships from document terms, and then, we aggregate them in a structured representation where domain information, and the related polarities, are preserved. Such a structured representation is stored in an index-like repository (from now simply referred as "index"). When the polarity of a new document has to be computed, its structured representation is built and, combined with domain information, it is used for querying the index in order to estimate the polarity of the whole document.

The rest of the work is structured as follows. Section 2 presents a survey on works about sentiment analysis. Section 3 described the proposed approach by explaining how texts are converted in a semantic structured representation, stored during the training phase, and exploited during the test one. Section 4 reports the comparison between the presented approach and three baselines. Finally, Sect. 6 concludes the paper.

2   Related Work

The topic of sentiment analysis has been studied extensively in the literature [2], where several techniques have been proposed and validated.

Machine learning techniques are the most common approaches used for addressing this problem, given that any existing supervised methods can be applied to sentiment classification. For instance, in [4], the authors compared the performance of Naive-Bayes, Maximum Entropy, and Support Vector Machines in sentiment analysis on different features like considering only unigrams, bigrams, combination of both, incorporating parts of speech and position information or by taking only adjectives. Moreover, beside the use of standard machine learning method, researchers have also proposed several custom techniques specifically for sentiment classification, like the use of adapted score function based on the evaluation of positive or negative words in product reviews [5], as well as by defining weighting schemata for enhancing classification accuracy [6].

An obstacle to research in this direction is the need of labeled training data, whose preparation is a time-consuming activity. Therefore, in order to reduce the labeling effort, opinion words have been used for training procedures. In [7,8], the authors used opinion words to label portions of informative examples for training the classifiers. Opinion words have been exploited also for improving the accuracy of sentiment classification, as presented in [9], where a framework incorporating lexical knowledge in supervised learning to enhance accuracy has been proposed. Opinion words have been used also for unsupervised learning approaches like the one presented in [10].

Another research direction concerns the exploitation of discourse-analysis techniques. [11] discusses some discourse-based supervised and unsupervised approaches for opinion analysis; while in [12], the authors present an approach to identify discourse relations.

The approaches presented above are applied at the document-level [13–18], i.e., the polarity value is assigned to the entire document content. However, in some case, for improving the accuracy of the sentiment classification, a more fine-grained analysis of a document is needed. Hence, the sentiment classification of the single sentences, has to be performed. In the literature, we may find approaches ranging from the use of fuzzy logic [19–23] to the use of aggregation techniques [24] for computing the score aggregation of opinion words. In the case of sentence-level sentiment classification, two different sub-tasks have to be addressed: (i) to determine if the sentence is subjective or objective, and (ii) in the case that the sentence is subjective, to determine if the opinion expressed in the sentence is positive, negative, or neutral. The task of classifying a sentence as subjective or objective, called "subjectivity classification", has been widely discussed in the literature [25–28] and systems implementing the capabilities of identifying opinion's holder, target, and polarity have been presented [29]. Once subjective sentences are identified, the same methods as for sentiment classification may be applied. For example, in [30] the authors consider gradable adjectives for sentiment spotting; while in [31–33] the authors built models to identify some specific types of opinions.

In the last years, with the growth of product reviews, the use of sentiment analysis techniques was the perfect floor for validating them in marketing activities [34–36]. However, the issue of improving the ability of detecting the different opinions concerning the same product expressed in the same review became a challenging problem. Such a task has been faced by introducing "aspect" extraction approaches that were able to extract, from each sentence, which is the aspect the opinion refers to. In the literature, many approaches have been proposed: conditional random fields (CRF) [37], hidden Markov models (HMM) [38], sequential rule mining [39], dependency tree kernels [40], clustering [41], neural networks [42,43], and genetic algorithms [44]. In [45,46], two methods were proposed to extract both opinion words and aspects simultaneously by exploiting some syntactic relations of opinion words and aspects.

A particular attention should be given also to the application of sentiment analysis in social networks [47–49]. More and more often, people use social networks for expressing their moods concerning their last purchase or, in general, about new products. Such a social network environment opened up new challenges due to the different ways people express their opinions, as described by [50,51], who mention "noisy data" as one of the biggest hurdles in analyzing social network texts.

One of the first studies on sentiment analysis on micro-blogging websites has been discussed in [52], where the authors present a distant supervision-based approach for sentiment classification.

At the same time, the social dimension of the Web opens up the opportunity to combine computer science and social sciences to better recognize, interpret, and process opinions and sentiments expressed over it. Such multi-disciplinary approach has been called *sentic computing* [53]. Application domains where sentic computing has already shown its potential are the cognitive-inspired classification of images [54], of texts in natural language, and of handwritten text [55].

Finally, an interesting recent research direction is domain adaptation, as it has been shown that sentiment classification is highly sensitive to the domain from which the training data is extracted. A classifier trained using opinionated documents from one domain often performs poorly when it is applied or tested on opinionated documents from another domain, as we demonstrated through the example presented in Sect. 1. The reason is that words and even language constructs used in different domains for expressing opinions can be quite different. To make matters worse, the same word in one domain may have positive connotations, but in another domain may have negative ones; therefore, domain adaptation is needed. In the literature, different approaches related to the Multi-Domain sentiment analysis have been proposed. Briefly, two main categories may be identified: (i) the transfer of learned classifiers across different domains [3,56,57], and (ii) the use of propagation of labels through graph structures [19,58–60].

All approaches presented above are based on the use of statistical techniques for building sentiment models. The exploitation of semantic information is not taken into account. In this work, we proposed a first version of a semantic-based approach preserving the semantic relationships between the terms of each sentence in order to exploit them either for building the model and for estimating document polarity. The proposed approach, falling into the multi-domain sentiment analysis category, instead of using pre-determined polarity information associated with terms, it learns them directly from domain-specific documents. Such documents are used for training the models used by the system.

3   The Approach

As introduced in Sect. 1, the proposed system is based on the implementation of an index-like approach, based on the use of structured representations of documents. Such representation is use for either preserving domain information associated with each document and for estimating the polarity of unclassified ones. Document polarity is estimated through the computation of a Score Status Value [61] (SSV) representing the aggregation of the polarities estimated for each feature extracted from the document. In this section, the steps carried out for implementing our approach are presented.

3.1   Feature Extraction

The first task consists in the detection of the features that are exploited for building the sentiment model. The proposed approach has been designed upon two main desiderata:

1. The need of preserving and exploiting semantic relationships between document terms, requires to find a structured representation of information able to address this issue. In particular, we want to store linguistic information of each term together with its semantic relationships with the other ones;
2. The described approach addresses the problem of sentiment analysis in a multi-domain environment; therefore, each extracted feature has to enclose domain-specific information in order to exploit them during the estimation of document polarity.

Addressing the two pillars described above, requires to parse raw texts in order to extract significant linguistic and semantic information. The proposed solution for extracting the set of features is based on the use of a native natural language processing library, namely the Stanford NLP Core Toolkit [62].

For each document of the training set, we applied the Stanford parser for extracting the terms dependencies. Such dependencies are taken into account for preserving the semantic between terms in the structured representation used for representing document content.

As an example, let's consider the following sentence:

"I came here to reflect my happiness by fishing."

By applying the Stanford parser, we obtain the following list of dependencies between terms:

```
nsubj(came-2, I-1)
nsubj(reflect-5, I-1)
root(ROOT-0, came-2)
advmod(came-2, here-3)
aux(reflect-5, to-4)
xcomp(came-2, reflect-5)
poss(happiness-7, my-6)
dobj(reflect-5, happiness-7)
prep_by(reflect-5, fishing-9)
```

Each dependency is composed by three elements: the name of the "relation" (R), the "governor" (G) that is the first term of the dependency, and the "dependent" (D) that is the second one. First of all, we removed from the dependencies list, ones containing a stop word[1] as governor or dependent element. Exceptions are made when one of the two terms contained in a dependency is an adjective. From the dependencies list presented above, the pruned list is the following:

```
poss(happiness-7, my-6)
dobj(reflect-5, happiness-7)
prep_by(reflect-5, fishing-9)
```

Then, for each dependency contained in the pruned list, we compile a set of pairs "field - value". Each pair is a "feature" associated with the dependency extracted from the document. Table 1 show, by using as example the dependency "dobj(reflect-5, happiness-7)", the list of extracted features.

Table 1. Field structure and corresponding content stored in the index.

Field name	Content
RGD	"dobj-reflect-happiness"
RDG	"dobj-happiness-reflect"
GD	"reflect-happiness"
DG	"happiness-reflect"
G	"reflect"
D	"happiness"

---

[1] The list of stop words used in this work is the one provided by Apache with the Lucene and Solr packages.

There are three considerations explaining the rationale of using the presented set of six features.

- The choice of considering the governor and the dependent in both orders is to meet the possibility that the parser may produce different output based on how the text is written within the sentence. Such an order is affected also by the parser used. In our approach we decided to adopt the Stanford parser, but, obviously, any parser producing a list of dependencies like the one presented above can be used.
- For the same reason, we decided to extract features pruned by the relation element, because different parsers may use different kind of dependencies. The meaning of these features (the third and fourth ones) is to track the co-occurrence of terms independently by the relationship between them.
- Finally, the "G" and "D" features are used as backup purpose. Indeed, if, for training a particular model, a small number of samples is available, the use of single terms allows to apply a bag-of-words approach as a backup for computing document polarity. For these two features only nouns, verbs, adverbs, or adjectives are considered.

The set of features extracted from each dependencies is given as input to the component that will combine such features with either the polarity and domain information in order to construct the final representation of each document.

3.2 Structured Representation Construction

Once all features have been extracted, they are passed to the component in charge of structuring and storing them in the model repository that, for simplicity, we call "index". As mentioned early, to each feature, the domain and polarity information are associated for building its equivalent structured representation. Where, the polarity associated with each feature contained in the model is the average of the polarities of the document in which each feature occurs. This shrewdness is necessary for distinguishing the polarities that each feature may assume in different domains. Indeed, classic approached based on the use of polarized vocabularies do not consider the possibility that a particular feature may assume different polarities depending on the context in which they occur. An example has been presented in Sect. 1.

On the light of this, the construction of the structured representation of each feature has to consider two aspects: (i) each feature may appear in different domains, and (ii) for each feature an estimation of the polarity for each domain has to be computed.

Therefore, each feature is translated into the correspondent structured representation shown below. By considering as example the feature "RGD - dobj-reflect-happiness", we have the following structure:

```
feature-type: RGD
feature-value: dobj-reflect-happiness
domain_1: polarity_1
```

```
domain_2: polarity_2
...
domain_n: polarity_n
```

The estimation of $polarity_i$ values associated with each domain is done by analyzing only the explicit information extracted from the training set. Values are computed as:

$$\text{polarity}_i(F) = \frac{k_F^i}{T_F^i} \in [-1, 1] \qquad \forall i = 1, \ldots, n, \tag{2}$$

where F is the feature taken into account, index i refers to domain D_i which the feature belongs to, n is the number of domains available in the training set, k_C^i is the arithmetic sum of the polarities observed for the feature F in the training set restricted to domain D_i, and T_C^i is the number of instances of the training set, restricted to domain D_i, in which feature F occurs.

Once all structured representation are built, they are stored in the repository. Such repository represents a multi-domain model for sentiment analysis purpose.

3.3 Polarity Computation

When an unclassified document needs to be evaluated, a procedure similar to the one adopted for building the model is used for computing its polarity.

A document is given as input to the Stanford parser and the list of dependencies is extracted and pruned by the ones containing stop words. Then, for each valid dependency, we build the related structured representation and we use it for estimating the polarity by analyzing information contained in the model. The final document polarity will be the average of the polarities estimated for each extracted dependency.

Let's consider the following sentence:

"I feel good and I feel healthy."

After the execution of the Stanford parser and the pruning of exceeding dependencies by using the same strategy described early, we obtain the following set of dependencies:

```
acomp(feel-2, good-3)
acomp(feel-6, healthy-7)
```

From these two dependencies, we generate the following two structures:

```
FEATURE ID: F1
feature-type: RGD; feature-value: acomp-feel-good
feature-type: RDG; feature-value: acomp-good-feel
feature-type: GD; feature-value: feel-good
feature-type: DG; feature-value: good-feel
feature-type: G; feature-value: feel
```

```
feature-type: D; feature-value: good

FEATURE ID: F2
feature-type: RGD; feature-value: acomp-feel-healthyd
feature-type: RDG; feature-value: acomp-healthy-feel
feature-type: GD; feature-value: feel-healthy
feature-type: DG; feature-value: healthy-feel
feature-type: G; feature-value: feel
feature-type: D; feature-value: healthy
```

For each structure I presented above, for which the domain D is given, we computed the SSV representing the polarity of the structure I in the domain which the structure belongs to. The Equation below, show how the SSV is computed.

$$
\begin{aligned}
SSV(I) = AVG(DP(RGD_{F1}) + DP(RDG_{F1}) + \\
DP(GD_{F1}) + DP(DG_{F1}) + \\
DP(G_{F1}) + DP(D_{F1}) + \\
DP(RGD_{F2}) + DP(RDG_{F2}) + \\
DP(GD_{F2}) + DP(DG_{F2}) + \\
DP(G_{F2}) + DP(D_{F2}))
\end{aligned}
\tag{3}
$$

where DP is the function extracting the polarity of the feature I for the domain D, and AVG refers to the averaging operation of all detected polarities.

4   Experimental Evaluation

In this Section, we present the results obtained from our experimental campaign where we compared our representation in different settings.

Dataset construction And Baselines The training and testing of the system has been done on two different dataset. For creating the training model, we built structured document representation by using reviews contained in the Blitzer dataset. In particular, we used the balanced version of the dataset in order to same number of positive and negative samples. Concerning the test operation, we created a test set of 32.000 reviews compiled by using the same strategy used for building the Blitzer dataset[2]. Test set is even balanced with respect to the number of positive and negative opinions. The same philosophy has been used for the domains, where, for each of the 16 domains used in the test set, we had 1.000 positive, and as many negative, reviews.

Our approach (Structured Domain Dependent, SDD) has been compared with three baselines:

- Most Frequent Polarity: the accuracy obtained by the system if it guesses the same polarity for all samples contained in the test set.

---

[2] The test set is available at https://goo.gl/siOJbZ.

- Structured Domain Independent: the accuracy obtained by using the proposed structured representation without considering domain information.
- Bag-Of-Word Domain Dependent: the accuracy obtained by using the classic statistical bag-of-words approach by considering also domain information.

Results and Discussion. Table 2 shows the results obtained by the three baselines and by the proposed approach. First column contains the name of the approach, while the second one the accuracy obtained on the test set.

Table 2. Accuracy obtained by our approach with respect to the three chosen baselines.

Approach	Accuracy
Most Frequent Polarity (MFP)	0.5000
Structured Domain Independent (SDI)	0.5407
Bag-Of-Word Domain Dependent (BDD)	0.6350
Structured Domain Dependent (SDD)	**0.6834**

Results show that the proposed approach leads to better results with respect to all the baselines. Beside this, there is also a significant difference between the accuracies obtained by using domain-dependent features (BDD and SDD approaches) and the one obtained without considering domain information.

By focusing on the two approaches exploiting domain information, in Table 3, we reported the detailed accuracy obtained on each domain by the two approaches exploiting such information. First column contain the name of the domain, second column the number of features for each domain and the last two columns the accuracies obtained by the BDD and SDD approaches respectively.

By observing the results reported in Table 3, no particular correlations between the number of features and the accuracy of the approach can be noticed. Unexpectedly, the worst result is obtained for the domain having the higher number of features, and one of the best results, obtained on the "tools_hardware" domain, is reported with a very low number of features compared to the others. One of the possible reasons may be the significant presence, in the set of documents used for building the model, of features having uncertain polarity, Indeed, if many features are used in either positive and negative contexts, it is difficult for the system to exploiting such information during the test phase for estimating document polarity. Further investigation in this direction may clarify this aspect.

Finally, we may notice that for the two domains, "gourmet_food" and "baby", the performance of the bag of words approach, outperform the semantic one.

5   ESWC-2018 SSA Challenge Tasks #1 and #2

Tables 4 and 5 shows the full results of Tasks #1 and #2 participants.
Approach Limits. As we mentioned at the end of Sect. 2, the approach presented in this paper is a first attempt of exploring the use of structured representation

Table 3. Accuracy obtained in each domain by the BDD and SDD approaches.

Domain	Features	BDD Accuracy	SDD Accuracy
Automotive	259,239	0.6230	0.6935
Baby	924,365	0.5980	0.5830
Beauty	601,163	0.6390	0.6470
Cell_phones_service	484,796	0.6115	0.6570
Computer_video_games	1,247,408	0.5165	0.5725
Electronics	944,796	0.6155	0.7180
Gourmet_food	417,309	0.6310	0.6275
Health_personal_care	768,616	0.6590	0.7180
Jewelry_watches	358,677	0.6375	0.6540
Kitchen_housewares	793,167	0.6460	0.7290
Musical_instruments	130,005	0.6540	0.7225
Office_products	180,172	0.6535	0.7105
Software	1,146,081	0.6680	0.7070
Sports_outdoors	869,576	0.6540	0.6810
Tools_hardware	40,962	0.6830	0.7250
Toys_games	833,887	0.6700	0.7885

Table 4. Results obtained by the participant to Task #1.

System	F-Measure
Guangyuan et al.	0.9643
Atzori et al.	0.9561
Federici et al.	0.9356
Dragoni et al.	0.9228
Dragoni et al.	**0.8823**
Rexha et al.	0.8743
Indurthi et al.	0.8203
Petrucci et al.	0.7153
Chaudhuri et al.	0.5243

Table 5. Results obtained by the participant to Task #2.

System	F-Measure
Federici et al.	0.5682
Dragoni et al.	**0.5244**
Rexha et al.	0.4598

of documents for addressing the sentiment analysis problem. For this reason, we performed a critical analysis of our work in order to highlight which are its limits and to outline a roadmap for future implementations. In particular, we detected three directions for extending the proposed approach:

- Improve dependencies pruning: in the feature extraction process, we pruned part of the dependencies extracted by the Stanford parser. In the light of the results reported in Table 3, we inferred that having a huge number of features is not preparatory for obtaining higher results. Therefore, a more restrictive policy should be implemented in pruning dependencies by trying to detect the most significant features despite the ones causing information overlapping between domains.
- Language coverage: a typical problem affecting the construction of language models is the language coverage of such models. Indeed, without having a large corpus for training the system, a significant number of terms information might be excluded. This issue is strictly connected with the next one and it may share the possible solution.
- Improve the semantic aspect: one of the possibility for addressing the problem of language coverage, is the adoption of external semantic resources, for instance WordNet, for extending the meaning of each feature. This way, we will be able to reduce the total number of features, due to the use of a concept-based representation of each feature instead of a term-based one, and, at the same time, to increase the language coverage. Working in this direction will mean that the current structured representation will have to be revised accordingly.

6   Conclusion

In this paper, we described a system exploiting a structured representation of document for the problem of multi-domain sentiment analysis. Even if the representation used for structuring documents and the metric adopted for estimating document polarity is quite simple, the system obtained reasonable performances in the provided evaluation. Future work will address the possibility to exploit more sophisticated metrics considering the belonging of a document to a certain domain not in a binary but in a *fuzzy* fashion, measuring some sort of semantic relatedness of the sentence under test with each domain and using such measures as weights for the polarity detection phase. Moreover, we intend to explore the integration of knowledge bases in order to move toward a more cognitive technique able to improve the language coverage of the approach.

References

1. Pang, B., Lee, L., Vaithyanathan, S.: Thumbs up? sentiment classification using machine learning techniques. In: Proceedings of EMNLP, Philadelphia, Association for Computational Linguistics, pp. 79–86 (July 2002)

2. Liu, B., Zhang, L.: A survey of opinion mining and sentiment analysis. In: Aggarwal, C.C., Zhai, C.X. (eds.) Mining Text Data, pp. 415–463. Springer, Berlin (2012)
3. Blitzer, J., Dredze, M., Pereira, F.: Biographies, bollywood, boom-boxes and blenders: domain adaptation for sentiment classification. In: ACL, pp. 187–205 (2007)
4. Pang, B., Lee, L.: A sentimental education: sentiment analysis using subjectivity summarization based on minimum cuts. In: ACL, pp. 271–278 (2004)
5. Dave, K., Lawrence, S., Pennock, D.M.: Mining the peanut gallery: opinion extraction and semantic classification of product reviews. In: WWW, pp. 519–528 (2003)
6. Paltoglou, G., Thelwall, M.: A study of information retrieval weighting schemes for sentiment analysis. In: ACL, pp. 1386–1395 (2010)
7. Tan, S., Wang, Y., Cheng, X.: Combining learn-based and lexicon-based techniques for sentiment detection without using labeled examples. In: SIGIR, pp. 743–744 (2008)
8. Qiu, L., Zhang, W., Hu, C., Zhao, K.: Selc: a self-supervised model for sentiment classification. In: CIKM, pp. 929–936 (2009)
9. Melville, P., Gryc, W., Lawrence, R.D.: Sentiment analysis of blogs by combining lexical knowledge with text classification. In: KDD, pp. 1275–1284 (2009)
10. Taboada, M., Brooke, J., Tofiloski, M., Voll, K.D., Stede, M.: Lexicon-based methods for sentiment analysis. Comput. Linguist. **37**(2), 267–307 (2011)
11. Somasundaran, S.: Discourse-level relations for Opinion Analysis. Ph.D. thesis, University of Pittsburgh (2010)
12. Wang, H., Zhou, G.: Topic-driven multi-document summarization. In: IALP, pp. 195–198 (2010)
13. Dragoni, M.: Shellfbk: an information retrieval-based system for multi-domain sentiment analysis. In: Proceedings of the 9th International Workshop on Semantic Evaluation. SemEval '2015, Denver, Colorado, pp. 502–509. Association for Computational Linguistics (June 2015)
14. Petrucci, G., Dragoni, M.: An information retrieval-based system for multi-domain sentiment analysis. In: Gandon, F., Cabrio, E., Stankovic, M., Zimmermann, A. (eds.) SemWebEval 2015. CCIS, vol. 548, pp. 234–243. Springer, Cham (2015). https://doi.org/10.1007/978-3-319-25518-7_20
15. Rexha, A., Kröll, M., Dragoni, M., Kern, R.: Exploiting propositions for opinion mining. In: Sack, H., Dietze, S., Tordai, A., Lange, C. (eds.) SemWebEval 2016. CCIS, vol. 641, pp. 121–125. Springer, Cham (2016). https://doi.org/10.1007/978-3-319-46565-4_9
16. Federici, M., Dragoni, M.: A knowledge-based approach for aspect-based opinion mining. In: Sack, H., Dietze, S., Tordai, A., Lange, C. (eds.) SemWebEval 2016. CCIS, vol. 641, pp. 141–152. Springer, Cham (2016). https://doi.org/10.1007/978-3-319-46565-4_11
17. Rexha, A., Kröll, M., Dragoni, M., Kern, R.: Opinion mining with a clause-based approach. In: Dragoni, M., Solanki, M., Blomqvist, E. (eds.) SemWebEval 2017. CCIS, vol. 769, pp. 166–175. Springer, Cham (2017). https://doi.org/10.1007/978-3-319-69146-6_15
18. Federici, M., Dragoni, M.: Aspect-based opinion mining using knowledge bases. In: Dragoni, M., Solanki, M., Blomqvist, E. (eds.) SemWebEval 2017. CCIS, vol. 769, pp. 133–147. Springer, Cham (2017). https://doi.org/10.1007/978-3-319-69146-6_13
19. Dragoni, M., Tettamanzi, A.G., da Costa Pereira, C.: Propagating and aggregating fuzzy polarities for concept-level sentiment analysis. Cogn. Comput. **7**(2), 186–197 (2015)

20. Dragoni, M., Tettamanzi, A.G.B., da Costa Pereira, C.: A fuzzy system for concept-level sentiment analysis. In: Presutti, V., Stankovic, M., Cambria, E., Cantador, I., Di Iorio, A., Di Noia, T., Lange, C., Reforgiato Recupero, D., Tordai, A. (eds.) SemWebEval 2014. CCIS, vol. 475, pp. 21–27. Springer, Cham (2014). https://doi.org/10.1007/978-3-319-12024-9_2

21. Petrucci, G., Dragoni, M.: The IRMUDOSA system at ESWC-2016 challenge on semantic sentiment analysis. In: Sack, H., Dietze, S., Tordai, A., Lange, C. (eds.) SemWebEval 2016. CCIS, vol. 641, pp. 126–140. Springer, Cham (2016). https://doi.org/10.1007/978-3-319-46565-4_10

22. Dragoni, M., Petrucci, G.: A fuzzy-based strategy for multi-domain sentiment analysis. Int. J. Approx. Reason. **93**, 59–73 (2018)

23. Petrucci, G., Dragoni, M.: The IRMUDOSA system at ESWC-2017 challenge on semantic sentiment analysis. In: Dragoni, M., Solanki, M., Blomqvist, E. (eds.) SemWebEval 2017. CCIS, vol. 769, pp. 148–165. Springer, Cham (2017). https://doi.org/10.1007/978-3-319-69146-6_14

24. da Costa Pereira, C., Dragoni, M., Pasi, G.: A prioritized "and" aggregation operator for multidimensional relevance assessment. In: Serra, R., Cucchiara, R. (eds.) AI*IA 2009. LNCS (LNAI), vol. 5883, pp. 72–81. Springer, Heidelberg (2009). https://doi.org/10.1007/978-3-642-10291-2_8

25. Federici, M., Dragoni, M.: Towards unsupervised approaches for aspects extraction. In: Dragoni, M., Recupero, D.R., Denecke, K., Deng, Y., Declerck, T. (eds.) Joint Proceedings of the 2th Workshop on Emotions, Modality, Sentiment Analysis and the Semantic Web and the 1st International Workshop on Extraction and Processing of Rich Semantics from Medical Texts co-located with ESWC 2016, Heraklion, Greece, May 29, 2016. Volume 1613 of CEUR Workshop Proceedings (2016). www.CEUR-WS.org

26. Federici, M., Dragoni, M.: A branching strategy for unsupervised aspect-based sentiment analysis. In: Dragoni, M., Recupero, D.R. (eds.)Proceedings of the 3rd International Workshop at ESWC on Emotions, Modality, Sentiment Analysis and the Semantic Web co-located with 14th ESWC 2017, Portroz, Slovenia, May 28, 2017. Volume 1874 of CEUR Workshop Proceedings (2017). www.CEUR-WS.org

27. Riloff, E., Patwardhan, S., Wiebe, J.: Feature subsumption for opinion analysis. In: EMNLP, pp. 440–448 (2006)

28. Wilson, T., Wiebe, J., Hwa, R.: Recognizing strong and weak opinion clauses. Comput. Intell. **22**(2), 73–99 (2006)

29. Palmero Aprosio, A., Corcoglioniti, F., Dragoni, M., Rospocher, M.: Supervised opinion frames detection with RAID. In: Gandon, F., Cabrio, E., Stankovic, M., Zimmermann, A. (eds.) SemWebEval 2015. CCIS, vol. 548, pp. 251–263. Springer, Cham (2015). https://doi.org/10.1007/978-3-319-25518-7_22

30. Hatzivassiloglou, V., Wiebe, J.: Effects of adjective orientation and gradability on sentence subjectivity. In: COLING, pp. 299–305 (2000)

31. Kim, S.M., Hovy, E.H.: Crystal: analyzing predictive opinions on the web. In: EMNLP-CoNLL, pp. 1056–1064 (2007)

32. Rexha, A., Kröll, M., Dragoni, M., Kern, R.: Polarity classification for target phrases in tweets: a Word2Vec approach. In: Sack, H., Rizzo, G., Steinmetz, N., Mladenić, D., Auer, S., Lange, C. (eds.) ESWC 2016. LNCS, vol. 9989, pp. 217–223. Springer, Cham (2016). https://doi.org/10.1007/978-3-319-47602-5_40

33. Rexha, A., Kröll, M., Kern, R., Dragoni, M.: An embedding approach for microblog polarity classification. In: Dragoni, M., Recupero, D.R. (eds.) Proceedings of the 3rd International Workshop on Emotions, Modality, Sentiment Analysis and the

Semantic Web co-located with 14th ESWC 2017, Portroz, Slovenia, May 28, 2017. Volume 1874 of CEUR Workshop Proceedings (2017). www.CEUR-WS.org

34. Recupero, D.R., Dragoni, M., Presutti, V.: ESWC 15 challenge on concept-level sentiment analysis. In: Gandon, F., Cabrio, E., Stankovic, M., Zimmermann, A. (eds.) SemWebEval 2015. CCIS, vol. 548, pp. 211–222. Springer, Cham (2015). https://doi.org/10.1007/978-3-319-25518-7_18

35. Dragoni, M., Reforgiato Recupero, D.: Challenge on fine-grained sentiment analysis within ESWC2016. In: Sack, H., Dietze, S., Tordai, A., Lange, C. (eds.) SemWebEval 2016. CCIS, vol. 641, pp. 79–94. Springer, Cham (2016). https://doi.org/10.1007/978-3-319-46565-4_6

36. Dragoni, M., Solanki, M., Blomqvist, E. (eds.): SemWebEval 2017. CCIS, vol. 769. Springer, Cham (2017). https://doi.org/10.1007/978-3-319-69146-6

37. Jakob, N., Gurevych, I.: Extracting opinion targets in a single and cross-domain setting with conditional random fields. In: EMNLP, pp. 1035–1045 (2010)

38. Jin, W., Ho, H.H., Srihari, R.K.: Opinionminer: a novel machine learning system for web opinion mining and extraction. In: KDD, pp. 1195–1204 (2009)

39. Liu, B., Hu, M., Cheng, J.: Opinion observer: analyzing and comparing opinions on the web. In: WWW, pp. 342–351 (2005)

40. Wu, Y., Zhang, Q., Huang, X., Wu, L.: Phrase dependency parsing for opinion mining. In: EMNLP, pp. 1533–1541 (2009)

41. Su, Q., Xu, X., Guo, H., Guo, Z., Wu, X., Zhang, X., Swen, B., Su, Z.: Hidden sentiment association in chinese web opinion mining. In: WWW, pp. 959–968(2008)

42. Dragoni, M.: NEUROSENT-PDI at semeval-2018 task 1: Leveraging a multi-domain sentiment model for inferring polarity in micro-blog text. In: Apidianaki, M., Mohammad, S.M., May, J., Shutova, E., Bethard, S., Carpuat, M. (eds.) Proceedings of The 12th International Workshop on Semantic Evaluation, SemEval@NAACL-HLT, New Orleans, Louisiana, June 5–6, 2018, pp. 102–108. Association for Computational Linguistics (2018)

43. Dragoni, M.: NEUROSENT-PDI at semeval-2018 task 3: Understanding irony in social networks through a multi-domain sentiment model. In: Apidianaki, M., Mohammad, S.M., May, J., Shutova, E., Bethard, S., Carpuat, M. (eds.) Proceedings of The 12th International Workshop on Semantic Evaluation, SemEval@NAACL-HLT, New Orleans, Louisiana, June 5–6, 2018, pp. 512–519. Association for Computational Linguistics (2018)

44. Dragoni, M., Azzini, A., Tettamanzi, A.G.B.: A novel similarity-based crossover for artificial neural network evolution. In: Schaefer, R., Cotta, C., Kołodziej, J., Rudolph, G. (eds.) PPSN 2010. LNCS, vol. 6238, pp. 344–353. Springer, Heidelberg (2010). https://doi.org/10.1007/978-3-642-15844-5_35

45. Qiu, G., Liu, B., Bu, J., Chen, C.: Opinion word expansion and target extraction through double propagation. Comput. Linguist. 37(1), 9–27 (2011)

46. Dragoni, M., da Costa Pereira, C., Tettamanzi, A.G.B., Villata, S.: Combining argumentation and aspect-based opinion mining: the smack system. AI Commun. 31(1), 75–95 (2018)

47. Dragoni, M.: A three-phase approach for exploiting opinion mining in computational advertising. IEEE Intell. Syst. 32(3), 21–27 (2017)

48. Dragoni, M., Petrucci, G.: A neural word embeddings approach for multi-domain sentiment analysis. IEEE Trans. Affect. Comput. 8(4), 457–470 (2017)

49. Dragoni, M.: Computational advertising in social networks: an opinion mining-based approach. In: Haddad, H.M., Wainwright, R.L., Chbeir, R. (eds.) Proceedings of the 33rd Annual ACM Symposium on Applied Computing, SAC 2018, Pau, France, April 09–13, 2018, pp. 1798–1804. ACM (2018)

50. Barbosa, L., Feng, J.: Robust sentiment detection on twitter from biased and noisy data. In: COLING (Posters), pp. 36–44 (2010)
51. Bermingham, A., Smeaton, A.F.: Classifying sentiment in microblogs: is brevity an advantage? In: CIKM, pp. 1833–1836 (2010)
52. Go, A., Bhayani, R., Huang, L.: Twitter sentiment classification using distant supervision. CS224N Project Report, Standford University (2009)
53. Cambria, E., Hussain, A.: Sentic computing: a common-sense-based framework for concept-level sentiment analysis (2015)
54. Cambria, E., Hussain, A.: Sentic album: Content-, concept-, and context-based online personal photo management system. Cogn. Comput. 4(4), 477–496 (2012)
55. Wang, Q.F., Cambria, E., Liu, C.L., Hussain, A.: Common sense knowledge for handwritten chinese recognition. Cogn. Comput. 5(2), 234–242 (2013)
56. Pan, S.J., Ni, X., Sun, J.T., Yang, Q., Chen, Z.: Cross-domain sentiment classification via spectral feature alignment. In: WWW, pp. 751–760 (2010)
57. Yoshida, Y., Hirao, T., Iwata, T., Nagata, M., Matsumoto, Y.: Transfer learning for multiple-domain sentiment analysis–identifying domain dependent/independent word polarity. In: AAAI, pp. 1286–1291 (2011)
58. Ponomareva, N., Thelwall, M.: Semi-supervised vs. cross-domain graphs for sentiment analysis. In: RANLP, pp. 571–578 (2013)
59. Huang, S., Niu, Z., Shi, C.: Automatic construction of domain-specific sentiment lexicon based on constrained label propagation. Knowl.-Based Syst. 56, 191–200 (2014)
60. Dragoni, M., da Costa Pereira, C., Tettamanzi, A.G.B., Villata, S.: Smack: an argumentation framework for opinion mining. In: Kambhampati, S. (ed.) Proceedings of the Twenty-Fifth International Joint Conference on Artificial Intelligence, IJCAI 2016, New York, NY, USA, 9–15 July 2016, pp. 4242–4243. IJCAI/AAAI Press (2016)
61. da Costa Pereira, C., Dragoni, M., Pasi, G.: Multidimensional relevance: prioritized aggregation in a personalized information retrieval setting. Inf. Process. Manag. 48(2), 340–357 (2012)
62. Manning, C.D., Surdeanu, M., Bauer, J., Finkel, J., Bethard, S.J., McClosky, D.: The Stanford CoreNLP natural language processing toolkit. In: Proceedings of 52nd Annual Meeting of the Association for Computational Linguistics: System Demonstrations, Baltimore, Maryland, pp. 55–60. Association for Computational Linguistics (June 2014)

Evaluating Quality of Word Embeddings with Sentiment Polarity Identification Task

Vijayasaradhi Indurthi[(⊠)] and Subba Reddy Oota

International Institute of Information Technology, Hyderabad, India
vijaya.saradhi@research.iiit.ac.in, oota.subba@students.iiit.ac.in

Abstract. Neural word embeddings have been widely used in modern NLP applications as they provide vector representation of words and capture the semantic properties of words and the linguistic relationship between the words. Many research groups have released their own version of word embeddings. However, they are trained on generic corpora, which limits their direct use for domain specific tasks. In this paper, we evaluate a set of pretrained word embeddings which were provided to us, on a standard NLP task - Sentiment Polarity Identification Task.

Keywords: Word-embeddings · NLP · Sentiment analysis

1 Introduction

Word embedding is a technique in Natural Language Processing which maps words of a language into dense vectors of real numbers in a continuous embedding space. Traditional NLP systems such as BoW(Bag of Words), TF-IDF represents words as indices in a vocabulary which were not able to capture the semantic relationships between words. Word embedding techniques have been gaining popularity in a range of NLP tasks like Sentiment analysis [12,22], Named Entity Recognition [7,19], Question Answering [17], etc.

We also compare the efficiency of the model with the standard word embeddings GloVe [16].

2 Related Work

As word embeddings are one of the most successful feature engineering techniques for a variety of NLP tasks, there exists recent work on evaluation of word embeddings in representing word semantics. Previous work focuses on evaluating word embeddings generated by different approaches. In [2], presented the first systematic evaluation of word embeddings generated by count models such as Distributional Semantics Composition Toolkit (DISSECT[1]), Continuous Bag of Words (CBOW[2]), Distributional Memory mode[3], and Collobert

---

[1] http://clic.cimec.unitn.it/composes/.
[2] https://code.google.com/p/word2vec/.
[3] http://clic.cimec.unitn.it/dm/.

© Springer Nature Switzerland AG 2018
D. Buscaldi et al. (Eds.): SemWebEval 2018, CCIS 927, pp. 232–237, 2018.
https://doi.org/10.1007/978-3-030-00072-1_18

and Weston mode[4]. These count models were tested on fourteen benchmarks in five categories: semantic relatedness (semantic similarity tested on Rubenstein and Goodenoughs dataset [18] and WordSim353 [3]), synonym detection (pair a target from 4 synonym candidates in TOEFL dataset [8]), concept categorization (group the nominal tasks into categories in Almuhareb-Poesio dataset [1]), selectional preferences (Ulrike Pados dataset [15]), and analogy (Mikolovs dataset [13]). From the above count models, they found that the word2vec model, CBOW, performed the best for almost all the tasks.

Similar to the above work, [20] trained the CBOW model of word2vec [13], C&W embeddings [4], Hellinger PCA [9], GloVe [16], TSCCA [5] and Sparse Random Projections [11] on a 2008 GloVe dump, and tested on the same fourteen datasets. They also found that CBOW outperformed other embeddings on 10 datasets. In addition to this intrinsic evaluation, they conducted extrinsic evaluation by using the embeddings as input features to two downstream tasks, namely noun phrase chunking and sentiment classification, and found the results of CBOW were also among the best. A similar intrinsic evaluation conducted in [6] and they additionally evaluated the skip-gram models of word2vec [13], CSLM word embeddings [21], dependency-based word embeddings [10], and combined word embeddings on NLP tasks (i.e., part-Of-speech tagging (POS), chunking, named entity recognition, mention detection) and linguistic tasks using Mikolovs dataset [13] and the WordSim353 dataset [3]. They trained these word embeddings on the Gigaword corpus composed of 4 billion words and found that the dependency-based word embeddings gave the best performance on the NLP tasks and combination of the embeddings yielded significant improvement. In [14], describes the evaluation of both syntactic and semantic properties of the word embeddings and that the tasks should be closer to real-word applications.

In this work, we provide a comparison of the quality of word embeddings trained separately from different resources. We evaluated word embeddings by applying them to sentiment polarity identification task on a standard dataset using standard machine learning classification algorithms.

3   Experiments

We evaluate the efficacy of word embeddings by using the word embeddings as features for the downstream NLP task of identifying the sentiment polarity of the given dataset. We formulate the problem of identifying the sentiment polarity as a text classification task. To generate the features of a sentence, we tokenize the sentence into tokens and compute the average of all the embeddings of the tokens of the sentence. We use two machine learning algorithms - Logistic Regression and SVM to train the models. We have been provided with 3 sets of embedding sizes and each embedding sizes we have 3 versions of embeddings trained on different lengths of the epochs. For every word embedding provided to us, we train the models on the training set provided to us. We make predictions on the test set provided to us. In addition to the word embeddings provided to us, we

---

[4] http://ronan.collobert.com/senna/.

also test the efficacy of the standard pretrained GloVe [16] word embeddings as a baseline model.

4   Evaluation Methods

For this challenge we have been provided 3 sets of embeddings with different dimensions and each embedding trained after a certain number of epochs. We have also been provided with a huge dataset of Amazon product reviews in different categories annotated with the sentiment polarity - positive or negative sentiments.

We evaluate the efficacy of the word embeddings on two aspects. One, the dimensionality of the embeddings. How the dimensionality of the word embeddings affect the classification performance. Two, the number of epochs the embeddings were trained for.

We run the classification task for every possible combination of word embedding dimension, number of trained epochs and for each category of the dataset. In addition we report the classification purpose on the standard word embeddings like Glove [16]. Table 2 shows the F1 score of the participating systems in Sentiment Analysis Task (Task1 of ESWC2017).

Table 1. Results of different word embeddings using Logistic Regression and SVM

Dimensions	Epochs	LR			SVM		
		Prec.	Recall	F1	Prec.	Recall	F1
128	15	0.8140	0.8267	0.8203	0.8075	0.8103	0.8089
128	30	0.8141	0.8264	0.8202	0.8066	0.8102	0.8084
128	50	0.8141	0.8267	0.8204	0.8056	0.8095	0.8075
256	15	0.8256	0.8381	0.8318	0.8101	0.8087	0.8094
256	30	0.8272	0.8392	0.8331	0.8082	0.8081	0.8082
256	50	0.8260	0.8384	0.8321	0.8093	0.8083	0.8088
512	15	0.8395	0.8515	0.8455	0.8028	0.7912	0.7969
512	30	0.8406	0.8517	0.8461	0.8020	0.7919	0.7969
512	50	0.8338	0.8493	0.8440	0.8009	0.7892	0.7950
300	GloVe6B	0.7855	0.7964	0.7909	0.7697	0.7738	0.7718
300	GloVe840B	0.7937	0.8130	0.8033	0.8013	0.7957	0.7985

5   Results

Table 1 shows the value of Precision, Recall and the F1 score of the each of the individual models trained using different word embeddings on the test set.

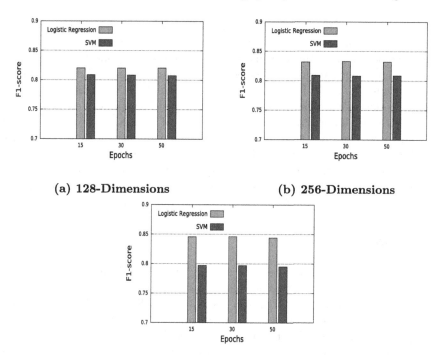

(a) 128-Dimensions   (b) 256-Dimensions

(c) 512-Dimensions

Fig. 1. Evaluation word embeddings with different dimensions (F1-score vs Number of epochs)

Table 2. F1-score for baseline systems in Sentiment Analysis Task

System	F-measure
Mattia Atzeni, Amna Dridi and Diego Reforgiato Recupero	0.8675
Fine-Grained Sentiment Analysis on Financial Microblogs and News Headlines [1]	
Marco Federici	0.8424
A Knowledge-based Approach For Aspect-Based Opinion Mining	
Walid Iguider and Diego Reforgiato Recupero	0.8378
Language Independent Sentiment Analysis of the Shukran Social Network using Apache Spark	
Giulio Petrucci	0.8112
The IRMUDOSA System at ESWC-2017 challenge on Semantic Sentiment Analysis	

An important observation is that both the models using Logistic Regression and Support Vector Machine for classification, using the given word embeddings have performed better than the baseline GloVe embeddings. Models using Logistic Regression in general performed better than the models using SVM. The number of dimensions was a key factor in the performance of the word embeddings. In general, models using higher dimension word embeddings performed better than the models using lower dimensions. The number of epochs the word embedding has trained on, does not seem to have a significant impact on the performance of the models. There is not much difference in the performance between 15 epoch trained word embeddings and 30 epoch trained word embeddings. In fact, training the neural network for 50 epochs had produced word embeddings which were slightly inferior to the ones which were trained for 30 epochs.

6   Future Work

In this paper we have evaluated the efficacy of the word embeddings using a downstream NLP task - Sentiment Polarity Identification, for which we used two machine learning algorithms - Logistic Regression and Support Vector Machines. We have used the plain version of the algorithms, without any hyper parameter tuning or regularization. In the future we would like to test the efficacy of the word embeddings with hyper parameter turning and regularization. Deep learning techniques are becoming popular in text classification. In the future we can examine the efficacy of the word embeddings by using deep learning models.

References

1. Almuhareb, A.: Attributes in lexical acquisition. Ph.D. thesis, University of Essex (2006)
2. Baroni, M., Dinu, G., Kruszewski, G.: Don't count, predict! a systematic comparison of context-counting vs. context-predicting semantic vectors. In: Proceedings of the 52nd Annual Meeting of the Association for Computational Linguistics (Volume 1: Long Papers), vol. 1, pp. 238–247 (2014)
3. Cinková, Silvie: WordSim353 for czech. In: Sojka, Petr, Horák, Aleš, Kopeček, Ivan, Pala, Karel (eds.) TSD 2016. LNCS (LNAI), vol. 9924, pp. 190–197. Springer, Cham (2016). https://doi.org/10.1007/978-3-319-45510-5_22
4. Collobert, R., Weston, J., Bottou, L., Karlen, M., Kavukcuoglu, K., Kuksa, P.: Natural language processing (almost) from scratch. J. Mach. Learn. Res. **12**(Aug), 2493–2537 (2011)
5. Dhillon, P., Rodu, J., Foster, D., Ungar, L.: Two step CCA: a new spectral method for estimating vector models of words. arXiv preprint arXiv:1206.6403 (2012)
6. Ghannay, S., Favre, B., Esteve, Y., Camelin, N.: Word embedding evaluation and combination
7. Lample, G., Ballesteros, M., Subramanian, S., Kawakami, K., Dyer, C.: Neural architectures for named entity recognition. arXiv preprint arXiv:1603.01360 (2016)
8. Landauer, T.K., Dumais, S.T.: A solution to plato's problem: the latent semantic analysis theory of acquisition, induction, and representation of knowledge. Psychol. Rev. **104**(2), 211 (1997)

9. Lebret, R., Collobert, R.: Word emdeddings through hellinger PCA. arXiv preprint arXiv:1312.5542 (2013)
10. Levy, O., Goldberg, Y.: Dependency-based word embeddings. In: Proceedings of the 52nd Annual Meeting of the Association for Computational Linguistics (Volume 2: Short Papers), vol. 2, pp. 302–308 (2014)
11. Li, P., Hastie, T.J., Church, K.W.: Very sparse random projections. In: Proceedings of the 12th ACM SIGKDD International Conference on Knowledge Discovery and Data Mining, pp. 287–296. ACM (2006)
12. Maas, A.L., Daly, R.E., Pham, P.T., Huang, D., Ng, A.Y., Potts, C.: Learning word vectors for sentiment analysis. In: Proceedings of the 49th Annual Meeting of the Association for Computational Linguistics: Human Language Technologies, vol. 1, pp. 142–150. Association for Computational Linguistics (2011)
13. Mikolov, T., Yih, W.t., Zweig, G.: Linguistic regularities in continuous space word representations. In: Proceedings of the 2013 Conference of the North American Chapter of the Association for Computational Linguistics: Human Language Technologies, pp. 746–751 (2013)
14. Nayak, N., Angeli, G., Manning, C.D.: Evaluating word embeddings using a representative suite of practical tasks. In: Proceedings of the 1st Workshop on Evaluating Vector-Space Representations for NLP, pp. 19–23 (2016)
15. Padó, S., Lapata, M.: Dependency-based construction of semantic space models. Comput. Linguist. **33**(2), 161–199 (2007)
16. Pennington, J., Socher, R., Manning, C.: Glove: Global vectors for word representation. In: Proceedings of the 2014 Conference on Empirical Methods in Natural Language Processing (EMNLP), pp. 1532–1543 (2014)
17. Ren, M., Kiros, R., Zemel, R.: Exploring models and data for image question answering. In: Advances in Neural Information Processing Systems, pp. 2953–2961 (2015)
18. Rubenstein, H., Goodenough, J.B.: Contextual correlates of synonymy. Commun. ACM **8**(10), 627–633 (1965)
19. Santos, C.N.d., Guimaraes, V.: Boosting named entity recognition with neural character embeddings. arXiv preprint arXiv:1505.05008 (2015)
20. Schnabel, T., Labutov, I., Mimno, D., Joachims, T.: Evaluation methods for unsupervised word embeddings. In: Proceedings of the 2015 Conference on Empirical Methods in Natural Language Processing, pp. 298–307 (2015)
21. Schwenk, H.: CSLM-a modular open-source continuous space language modeling toolkit (2013)
22. Tang, D., Wei, F., Yang, N., Zhou, M., Liu, T., Qin, B.: Learning sentiment-specific word embedding for twitter sentiment classification. In: Proceedings of the 52nd Annual Meeting of the Association for Computational Linguistics (Volume 1: Long Papers), vol. 1, pp. 1555–1565 (2014)

Author Index

Printed in the United States
By Bookmasters